I0031898

David O´Brian

Laboratory Guide

David O´Brian

Laboratory Guide

ISBN/EAN: 9783742862952

Manufactured in Europe, USA, Canada, Australia, Japa

Cover: Foto ©berggeist007 / pixelio.de

Manufactured and distributed by brebook publishing software
(www.brebook.com)

David O'Brian

Laboratory Guide

THE PRACTICAL

LABORATORY

Guide in Chemistry,

BY

DAVID O'BRINE, B. S., M. E.,

Assistant in Chemistry in the Ohio State University.

"*Longum iter est per praecepta, breve et efficax per exempla.*" — Seneca, Ep. vi.

COLUMBUS, O.,

A. H. SMYTHE,

1883.

Entered according to act of Congress, in the year 1883, by

DAVID O'BRINE,

In the office of the Librarian of Congress, at Washington.

PREFACE.

This little volume is intended for the use of students who possess some knowledge of Chemistry. The object is to present a practical guide in Chemistry adapted to the wants of the College or the Medical Laboratory. It would be impossible to acknowledge the sources of all analytical details or methods; they have been used, and in most cases modified by so many different persons that they are now regarded as common property. The labors of many well known chemists have been laid under contribution. Some of the methods are my own, *and every test presented has been verified.* There is a discussion of all that is important in the analysis of water, milk and cheese, blood, urine, and poisons. Especial attention is invited to the following subjects: Separation of Bases and tests; Comparison of Phosphorous, Arsenic, and Antimony; The Organic Acids; Classification of the Alkaloids; The Ptomaines; and Stoichiometry.

My thanks are due to the students and faculty of the Columbus Medical College and especially to Drs. HAMILTON, KINSMAN, LEE, and POOLEY, for their kindness and assistance. I am under special obligations to Professor NORTON, LL.D., of the Ohio State University, for much kind advice and assistance with the proof and many of the tests.

The work is presented to the Laboratory student, hoping it may lessen his labor, and also be of interest to the physician and general chemist.

OHIO STATE UNIVERSITY, Columbus, May, 1883.

ERRATTA.

Before the student uses this book, let him make the following corrections :

crystalizod read crystallized, on pages 3, 4, and 5.
CaO read CuO, on 6 page, article 23.
mercuric oxide read mercuric sulphate on page 8, article 37.
mercuric oxide read mercurous oxide, on page 9, article 39.
crucibel read crucible, on page 13, article 51.
disk read dish, on page 31.
instanae read instance, on page 34, foot note.
$K_4Fe_2Cy_6$ read K_4FeCy_6, on page 61, Zettnow's chart.
mercuric nitrate read mercuric, etc., on page 85, foot note.
crupreous urate read cuprous, etc., on page 87.
7,000 grains read 70,000 grains, on page 100.
looses read loses, on page 101.
buretts read burettes, on page 104, foot note.
pepetts read pipettes, on page 104, foot note.
us read as, on page 105, foot note.
sodum read sodium, on page 108, foot note.
nitrate read nitrite, on page 108, foot note, No. 1.
Na_2O_3 read N_2O_3, on page 108, foot note, No. 1.
.01 gram read .01 milligram, on page 108, foot note, No. 1.
analases read analysis, on page 114, under cheese.
physiloogical read physiological, on page 116.
destribution read distribution, on page 117, near the bottom.
heals read heels, on page 21, last line.
linnen read linen, on page 122.
strong alcohol read strong sulphuric acid, on page 128.
1,300 : 1,000 read 1,390 : 1,000, on page 175.

CONTENTS.

PART I.

	PAGE.
How the Reagents are Made	1–17
Definitions	18–22
Tests in the Dry Way	22–23
Separation of Groups —	
I. Pb, Ag, Hg	33–36
II. Hg, Pb, Cu, Cd, and As, Sb, Sn	36–44
III. Fe, Al, Cr, and Co, Ni, Mn, Zn	45–52
IV. Ba, Sr, Ca, Mg?	53–58
V. K, Na, NH_4HO	58–60
Zettnow's Separation without H_2S, or $(NH_4)_2S$	61–62
Acids — Dry Way	63–64
Wet Way { Mineral	64–73
{ Organic	73–77

PART II.

Urine	78
Urea	83
Uric Acid	85
Albumen	89
Chlorides	90
Phosphates	91
Grape Sugar	92
Sulphates	94
Bile	95

SYSTEMATIC ANALYSIS OF URINE...................... 96
    "   "   CALCULI 98

PART III.

WATER............   100
    Chlorine..............   103
    Ammonia, Free.......................   104
        "   Albuminoid.........................   104
    Hardness...................   106
    Total Solids................................   106
    Nitrates — Nitrites, sulphates   106

PART IV.

MILK............   110
    Water ..   112
    Fat   112
    Caseine   112
    Milk Sugar.........................   113
    Ash...................... ...   113
    Cheese   114

PART V.

POISONS —
    Ptomaines 116
    Alkaloids {Vegetable   118
              {Animal   119
    Systematic Separation.............................. 120
    Strychnine .   125
    Brucine   127
    Morphene   127
    Codeine   129
    Narcotine...............   129
    Quinine   130
    Cinchonine   131

Poisons (continued) —
    Veratrine 131
    Aconitine 132
    Atropine 133
    Nicotine 134
    Conine 135
    Caffeine and Theine 137
    Reagents for the Alkaloids 138
    Hydrocyanic acid 139
    Antimony 142
    Phosphorous 143
    Mercury 144
    Arsenic 146
    Lead 150
    Zinc 152
    Copper 153
    How they destroy Life 154

PART VI.

Blood 156
    Tests 158
Plant Bases 161
Incompatibility 163
Analysis of Man 165
    Daily Supply 166
    "   Waste 167

PART VII.

Stoichiometry 168–176
Index 177

LABORATORY GUIDE.

CHAPTER I.

PREPARATION OF REAGENTS.

ACIDS.

I. Acetic ($HC_2H_3O_2$), sp. gr. 1.04, contains 30 % acid.

(a) By treating alcohol with chromic acid to oxidize as follows: $C_2H_6O + O_2 = HC_2H_3O_2 + H_2O$.

(b) By treating an acetate with sulphuric acid as follows: $2(KC_2H_3O_2) + H_2SO_4 = K_2SO_4 + 2(\overline{HC_2H_3O_2})$.*

2. Hydrochloric (HCl), sp. gr. 1.12, contains 24 % acid.

(a) $2NaCl + H_2SO_4 = Na_2SO_4 + 2(\overline{HCl})$, by treating fused common salt with sulphuric acid.

(b) $2(H_2O) + 2(Cl_2) = 4HCl + \overline{O_2}$, by passing steam and chlorine through red hot porcelain tube.

3. Hydrosulphuric (H_2S), water saturated with the acid, made as required by treating ferrous sulphide with sulphuric acid.

$FeS + H_2SO_4 = FeSO_4 + \overline{H_2S}$. By the direct union of hydrogen and sulpher by heating them together, $S + H_2 = H_2S$.†

4. Hydrofluosilicic ($HF)_2SiF_4$. By treating powdered glass and fluor spar with sulphuric acid, $CaF_2 + H_2SO_4 = CaSO_4 +$

---

*In this book, the sign ‾‾‾‾ placed over a symbol indicates that a gas is produced or evolved; the same sign ____ placed under a symbol, that a solid is formed and precipitates.

†FeS is made on a small scale, by heating iron white hot in a blacksmith's forge, and running it through sulphur and collecting the product in water; dry and it is ready for use, $Fe + S = FeS$.

$\overline{(HF)}_2$, silicon can take the place of hydrogen, SiF_4, gaseous fluoride of silicon, $SiF_4 + 2(H_2O) = SiO_2 + 4(HF)$, this combines with a second portion to form $2(HF)SiF_4$ which does not corrode glass.*

5. Nitric (HNO_3), sp. gr. 1.2, contains 32 % acid. By heating a nitrate with sulphuric acid, $2(KNO_3) + H_2SO_4 = K_2SO_4 + 2(\overline{HNO_3})$. By treating nitric peroxide with water at ordinary temperatures, $3(N_2O_4) + 2(H_2O) = 4(HNO_3) + N_2O_2$.

6. Nitro-hydrochloric, three parts of HCl and one of HNO_3.

7. Nitrophenic, $HC_6H_2(NO_2)_3O$, "picric acid," made by the action of nitric acid and heat on phenol (C_6H_6O). The (NO_2) of HNO_3 replaces the H, making mon, di, or triphenol. This is a est for nitrates (*and explosive*) and nitrites, a few drops being sufficient as commonly made in the laboratory. Take one part of phenol (cryst. carbolic acid), four parts of strong sulphuric acid and two parts of water, the addition of potassium hydrate deepens and brightens the color.

8. Oxalic ($H_2C_2O_4$) crystals dissolved in ten parts of water.

(*a*) By the action of nitric acid on sugar or sawdust, $C_{12}H_{22}O_{11} + 18(O) = 6(C_2H_2O_4) + 5(H_2O)$.

(*b*) By the direct combination of an alkaline metal with carbonic anhydride, $2(CO_2) + 2Na = Na_2C_2O_4$.

(*c*) By the decomposition of cyanogen with water, $2(CN) + 4(H_2O) = C_2H_2O_4 + 2(\overline{NH_3})$.

Commercially made (*a*) by sawdust, which yields about one-half its weight of crystallized oxalic acid.

9. Sulphuric (H_2SO_4). The concentrated has sp. gr. 1.843, the ($H_2S_2O_7$) or [$H_2O, 2(SO_3)$], sp. gr. 1.9.

(*a*) Sulphurous acid is formed by burning S in air or by roasting pyrites, ($FeS_2 = FeS + \overline{S}$), $S + O_2 = SO_2$.

---

*1. $SiO_2 + 2(CaF_2) + 2(H_2SO_4) = SiF_4 + 2(CaSO_4) + 2(H_2O)$.
2. $3(SiF_4) + 2(H_2O) = 2(H_2SiF_6) + SiO_2$.

(b) The sulphurous acid is oxidized by nitric acid, nitric peroxide being set free (the nitric acid is made by treating sodic nitrate with sulphuric acid), $SO_2 + 2(HNO_3) = H_2SO_4 + N_2O_4$. In the presence of steam the N_2O_4 becomes nitric acid (HNO_3) and nitric oxide (N_2O_2), $3(N_2O_4) + 2(H_2O) = 4(HNO_3) + N_2O_2$. The nitric acid oxidizes more sulphurous anhydride (SO_2), while the N_2O_2 takes oxygen from the air and becomes N_2O_3 and N_2O_4, which also oxidizes sulphurous anhydride (SO_2), $N_2O_3 + SO_2 + H_2O = H_2SO_4 + N_2O_2$, when the operation becomes continuous. The chamber acid is concentrated in leaden pans until a sp. gr. 1.72 (commercial acid), when it is distilled in platinum retorts; these, though costing $8,000 to $10,000, are more profitable than glass ones.

10. Tartaric ($H_2C_4H_4O_6$) crystals dissolved in three parts of water, make as needed.

(a) Argols or tartar ($KHC_4H_4O_6$) deposited from fermenting grape juice is purified by subsequent crystallization (cream of tartar).

(b) This is dissolved in hot water and boiled with powdered chalk, forming an *insoluble* calcic tartrate and a *soluble* potassic tartrate, $2KHC_4H_4O_6 + CaCO_3 = K_2C_4H_4O_6 + CaC_4H_4O_6 + H_2O + CO_2$.

(c) Calcium chloride is now added to the clear solution, when all the tartaric acid is precipitated as calcic tartrate, $K_2C_4H_4O_6 + CaCl_2 = 2KCl + CaC_4H_4O_6$. This tartrate of lime is collected and boiled with dilute sulphuric acid, when an insoluble calcic sulphate is formed, the solution containing free tartaric that can be crystalized, $CaC_4H_4O_6 + H_2SO_4 = C_4H_6O_6 + CaSO_4$.

11. Bromine water (Br), water saturated with bromine, one part of bromine is soluble in thirty-four parts of water, sp. gr. 1.024.

12. Chlorine water (Cl), water saturated with chlorine, water absorbs at 10° C., 2.58 times its volume of chlorine; keep in a cool place in a bottle covered with black paper. The reactions for the halogen group:

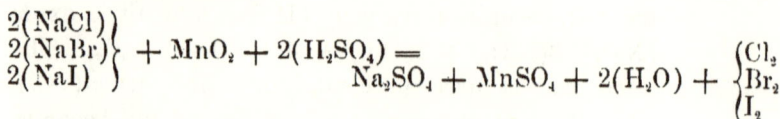

$$2(NaCl) \atop 2(NaBr) \atop 2(NaI) \} + MnO_2 + 2(H_2SO_4) = \atop Na_2SO_4 + MnSO_4 + 2(H_2O) + {Cl_2 \atop Br_2 \atop I_2}$$

13. Alcohol (C_2H_6O), sp. gr. .815, about 95 %.

1.   (a) By the action of hypochlorous acid on olefine (ethylene, C_2H_4), $C_2H_4 + HClO = C_2H_4Cl(OH)$, a chlorinated ethyl alcohol.

     (b) When this is acted upon by nascent hydrogen, $C_2H_4Cl(OH) + H_2 = C_2H_5(OH) + HCl$. *Note*, the ethylene may be prepared by the direct union of carbon and hydrogen.

2.   By the fermentation of grape sugar, $C_6H_{12}O_6 = 2(C_2H_6O) + 2(CO_2)$. There are other products formed, but these are the most important.

14. Absolute alcohol is formed by treating the strongest commercial alcohol with quick lime a number of times, and afterwards with *anhydrous* copper sulphate to remove the last traces of water.

15. Ammonium chloride (NH_4Cl), one part of crystallized salt in eight parts of water. When nitrogenous bodies decay, or when horns, bones, coal, etc., are subjected to destructive distillation, an impure ammonium carbonate is formed, a bye product from the gas works; when this is treated with hydrochloric acid, ammonium chloride is formed, $(NH_4)_2CO_3 + 2(HCl) = 2(NH_4)Cl + H_2CO_3$; this is purified by heating to dryness, and recrystalized.

16. Ammonium carbonate [$(NH_4)_2CO_3$], one part of the salt with four water and one part of ammonia; when used as a solvent for arsenious sulphide, the ammonia is omitted, $(NH_4)_4H_2(CO_3)_3$.

17. Ammonium hydrate (NH_4HO), sp. gr. .96, contains 10 % NH_3. Ammonium chloride is heated with slaked lime, and the gas passed into cold water, $2(NH_4Cl) + CaO = 2NH_3 + CaCl_2 + H_2O$. $H_2O + NH_3 = NH_4HO$.

18. Ammonium molybdate [$(NH_4)_2MoO_4$], a solution in nitric acid. The ores, molybdenite (MoS_2) and plumbic molybdate ($PbMoO_4$), especially the first, are heated in the air at a dull red heat, when $\overline{SO_2}$ and molybdic anhydride (MoO_3) and iron oxide are left. The residue is digested with strong ammonia, which dissolves the molybdic anhydride in the form of molybdate of ammonia, yielding prismatic crystals on evaporation.

19. Ammonium sulphide ($NH_4)_2S$, colorless; ($NH_4)_2S_2$ and NH_4HS, yellow. Pass H_2S through NH_4HO until it does not produce a precipitate in a solution of $MgSO_4$. It should not give a precipitate with a solution of lime, and no residue when evaporated to dryness and ignited.

*20. Ammonium oxalate, ($NH_4)_2C_2O_4$. Dissolve one part of the salt in twenty-four of water. Neutralize a solution of oxalic acid with ammonia, evaporate and crystalize. Should not be rendered turbid by H_2S or ($NH_4)_2S$, leaves no residue when ignited on Pt foil.

21. Barium chloride ($BaCl_2$), one part of crystalized salt in ten parts of water. The carbonate (witherite) or the sulphate (heavy spar) is roasted with charcoal, converting the carbonate into the oxide, and the sulphate into the soluble sulphide, $BaSO_4 + 4C = \overline{4CO} + BaS$. The crude product is dissolved in hydrochloric acid, as follows:

$BaS + 2HCl = \overline{H_2S} + BaCl_2$, dissolved and recrystallized in rhombic non-deliquescent plates.

22. Barium carbonate ($BaCO_3$). Dissolve $BaCl_2$ in hot water, and precipitating by ($NH_4)_2CO_3$ and NH_4HO, the precipi-

---

* Ammonium nitrate (NH_4NO_3), neutralize HNO_3 with ($NH_4)_2CO_3$; evaporate until crystals commence to form, then let cool. The crystals are afterwards fused.

tate must be washed until it gives no reaction with $AgNO_3$. Now make this to the consistency of cream with water, and keep in a stoppered bottle — shake before using.

23. Barium hydrate, $Ba(OH)_2$, a saturated watery solution, made by dissolving a solution of the soluble salts $(BaCl_2)$ in a boiling solution of NaHO, $BaCl_2 + (NaHO)_2 = Ba(OH)_2 + 2$ (NaCl), or by heating a solution of BaS with CaO.

24. Barium nitrate $[Ba(NO_3)_2]$, one part to fifteen of water. If the BaS (of 21) is dissolved in HNO_3, a nitrate is formed, $BaS + 2(HNO_3) = \overline{H_2S} + Ba(NO_3)_2$, dissolve and recrystallize.

25. Calcium chloride $(CaCl_2)$, one part of the salt in eight of water. Pure marble is treated with hydrochloric acid. The traces of silica, iron and magnesia, if present, are removed. $CaCO_3 + (HCl)_2 = CaCl_2 + \overline{H_2CO_3}$; this is dissolved and crystallized.

26. Calcium hydrate $Ca(OH)_2$, the dry solid and saturated watery solution. The same as [No. 23].

27. Calcium sulphate $(CaSO_4)$, a saturated watery solution, one part in 400 water, when finely pulverized (gypsum).

28. Carbon disulphide (CS_2), made (1) by passing the vapor of sulphur over heated charcoal, $C + S_2 = CS_2$; (2) by heating together charcoal and pyrites, $C + (FeS_2)_2 = CS_2 + (FeS)_2$.

29. Cobaltous nitrate $[CO(NO_3)_2]$, one part of the salt to eight of water. The oxide, freed from impurities, is dissolved in nitric acid.

30. Copper sulphate $(CuSO_4)$, one part of salt to eight parts water. In the manufacture of copper, the sulphide is roasted and oxidized to the sulphate; it can now be crystallized as commercial blue vitrol, or it may be precipitated with iron as metallic copper; this can be dissolved in sulphuric acid, but it invariably contains traces of iron: to purify it, oxidize the ferrous to the ferric sulphate of iron by a few drops of nitric acid; on concentrating, the liquid crystals of pure $CuSO_4$ are easily obtained.*

31. Ether $[(C_2H_5)_2O]$, sp. gr. not over .728, should not contain over 5 % of alcohol. Made by treating alcohol with strong sulphuric acid and distilling over into a receiver, $(C_2H_6O)_2 + H_2SO_4 = C_4H_{10}O + H_2O$.

32. Ferrous sulphate $(FeSO_4)$, one part of salt to ten of water, made as it is required for use. Take the ferrous sulphate, residues in making H_2S (No. 3), concentrate and filter, let it crystallize under a layer of alcohol ($\frac{1}{2}$ inch is enough of alcohol) —dry.

33. Ferric chloride (Fe_2Cl_6), one part salt to fifteen parts of water, should be neutral. Dissolve the iron or $Fe_2(OH)_6$ in hydrochloric acid; add a few drops of nitric acid to oxidize.

34. Gold chloride $(AuCl_3)$. The gold is precipitated with oxalic acid and dissolved in "aqua regia," avoiding an excess. Evaporate on the water bath to dryness. Dissolve in water; it should be neutral.†

35. Lead acetate $[Pb(C_2H_3O_2)_2]$, one part of the salt in ten of water. Dissolve the oxide in acetic acid, $PbO + (HC_2H_3O_2)_2 = Pb(C_2H_3O_2)_2 + H_2O$, using an excess of acetic acid to prevent basic salts forming.

36. Magnesium sulphate $(MgSO_4)$, one of the salt to ten of water.

---

*The *ammonio sulphate of copper* is made by adding NH_4HO drop by drop to a moderately concentrated solution of $CuSO_4$ until the precipitate at first produced is nearly redissolved; use the clear solution. The *ammonio nitrate of silver* is prepared in the same way, $AgNO_3$ taking the place of $CuSO_4$.

†Platinum chloride $(PtCl_4)$ is made in the same way as gold chloride. It is best to dissolve in a beaker flask.

(*1*) Made from the native carbonate by treating it with sulphuric acid, $MgCO_3 + H_2SO_4 = MgSO_4 + \overline{H_2CO_3}$. The carbonate ($CaMg2CO_3$) contains Ca, but it forms with H_2SO_4 an *insoluble*$CaSO_4$, while the $MgSO_4$ is *very soluble*, they can be readily separated in this way.

(*2*) From the "bittern" of sea water, which remains after the NaCl has crystallized out.

"Magnesium mixture", contains in a liter 101.5 grams of crystallized magnesium chloride, 200 grams of ammonium chloride, and 400 grams of ammonium hydrate (sp. gr. .96).

Millon's reagent. See alkaloids.

37. Mercuric chloride ($HgCl_2$), one part of the salt in sixteen of water.

(*1*) Dissolve mercury in aqua regia, with an excess of hydrochloric acid.

(*2*) By burning Hg in an excess of Cl gas.

(*3*) By dissolving mercuric oxide in hydrochloric acid.

(*4*) By treating mercuric oxide with common salt, $HgSO_4 + 2(NaCl) = Na_2SO_4 + HgCl_2$.

38. Mercurous chloride (Hg_2Cl_2), insoluble in water.

(*1*) By precipitating a solution of mercurous nitrate with hydrochloric acid or sodic chloride, $2(HgNO_3) + 2(HCl) = Hg_2Cl_2 + 2(HNO_3)$.

(*2*) By heating a mixture of corrosive sublimate (seventeen parts) with mercury (thirteen parts), $HgCl_2 + Hg = Hg_2Cl_2$.

(*3*) By heating (after they are well mixed) two parts $HgSO_4$ and three NaCl and four parts of Hg, $HgSO_4 + Hg + 2(NaCl) = Hg_2Cl_2 + Na_2SO_4$.

(*4*) By passing SO_2 through a saturated solution of crystalline mercuric chloride heated to 50° C. [122° F.], $2(HgCl_2) + 2(H_2O) + SO_2 = Hg_2Cl_2 + 2(HCl) + H_2SO_4$.

39. Mercurous nitrate ($HgNO_3$), **mercuric nitrate**, $Hg(NO_3)_2$. By treating Hg with a *slight excess* of dilute nitric acid, forms chiefly normal mercurous nitrate; ' when the mercury is in excess, basic mercurous nitrate is formed, as follows:

$[3(Hg_2)''2(NO_3), Hg_2''O, H_2O]$, basic nitrate. It can be known from the neutral salt by becoming black when triturated in a mortar with NaCl, calomel being formed, and mercuric oxide separated, as follows:

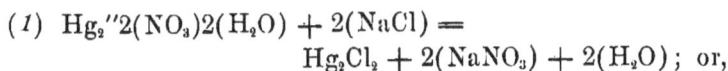

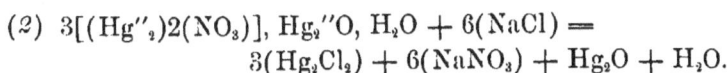

(1) $Hg_2''2(NO_3)2(H_2O) + 2(NaCl) =$
$\quad Hg_2Cl_2 + 2(NaNO_3) + 2(H_2O)$; or,

(2) $3[(Hg''_2)2(NO_3)], Hg_2''O, H_2O + 6(NaCl) =$
$\quad 3(Hg_2Cl_2) + 6(NaNO_3) + Hg_2O + H_2O.$

Hot dilute nitric acid in excess forms chiefly mercuric nitrate. Fuming nitric acid or nitrous acid hastens the solution. When the mercury is in excess, both basic mercurous and basic mercuric nitrate are formed. In all cases, chiefly nitric oxide gas is liberated.

40. Potassium chromate (K_2CrO_4), one part of the salt in ten parts of water.

(1) By adding potassic carbonate to a solution of potassic bichromate.

(2) By fusing a chromium compound with potassic carbonate.

(3) By heating powdered chrome ironstone with chalk and potassic carbonate in air — KNO_3 is sometimes added to hasten oxidation — the ferrous oxide is changed into insoluble ferric oxide, and the Cr_2O_3 into $2(CrO_3)$, which forms with the potash, K_2CrO_4, which can be dissolved out and purified by recrystallization.

41. Potassium bichromate ($K_2Cr_2O_7$), one part in ten parts of water. The potassic chromate (see 40) is oxidized by nitric acid, $2(K_2OCrO_3) + H_2ON_2O_5 = H_2O + K_2ON_2O_5 + K_2O2(CrO_3).$

If too much nitric acid is used, a ter-chromate is formed, K_2O $3(CrO_3)$.

42. Potassium chlorate ($KClO_3$), the crystallized salt. Pass a slow current of Cl into a cold dilute solution of KHO; potassium chloride and potassium hypochlorite are formed, $2(K_2OH_2O) + 4Cl = K_2OCl_2O + 2(KCl) + (H_2O)_2$. If this solution is boiled, it is converted into chlorate and chloride (best shown by reversing the equation), $3(K_2OCl_2O) = 4(KCl) + K_2OCl_2O_5$, or $2(KClO_3)$.

43. Potassium cyanide (KCN or KCy), one part of the salt to four of water. See prussic acid.

44. Potassium ferrocyanide (K_4FeCy_6), one part of the salt to twelve of water. By heating animal matter, horns, blood, leather, etc., with iron filings and K_2CO_3. The organic matter contains N and an excess of carbon. The C and N combine to form CN; now the excess of carbon reduces the K_2CO_3 to $K_2(?)$, which combines with CN to form KCN, and is converted by the presence of Fe into ferrocyanide.

45. Potassium ferricyanide (K_3FeCy_6), one part of the salt to twelve of water. The solution will not keep; make as it is wanted for use. It becomes reduced to the ferrocyanide in solution on standing. Oxidize the ferrocyanide by passing Cl through the cold dilute solution until it turns red. This salt is red and the ferro is yellow. It is decomposed by an excess of Cl and of H_2S.

46. Potassium iodide (KI), one part of the salt to twenty of water.

(*1*) Iodine is added to a solution of caustic potash, forming an iodide and an iodate, $6(KHO) + 6I = 5KI + KIO_3 + 3(H_2O)$. When KIO_3 is heated, $\overline{O_3}$ as KIO_3 = $KI + \overline{O_3}$.

(*2*) By dissolving potassic carbonate in hydriodic acid, $K_2CO_3 + 2(HI) = 2(KI) + CO_2, H_2O$.

(3) By digesting iodine and iron in water, ferric and ferrous iodides are formed (FeI_2 and Fe_2I_6), boil these salts and add K_2CO_3, avoiding an excess, when $FeI_2 + Fe_2I_6 + 4(K_2CO_3) = 8(KI) + Fe_3O_4 + 4(CO_2)$. The Fe_3O_4 is filtered off, and the solution of KI evaporated down.

(4) By adding K_2SO_4 to BaI_2 or CaI_2, as follows: $K_2SO_4 + BaI_2 = BaSO_4 + 2(KI)$, a white crystalline (cubic) solid; its solution dissolves iodine. It is decomposed by HNO_3, Cl, etc. The iodate may be known by giving off oxygen and by turning brown on the addition of HCl; effervescence indicates a carbonate.

47. Potassium mercuric iodide (Nessler's reagent). See water analysis.

48. Potassium metanimoniate ($KSbO_3$). See test for sodium.

49. Potassium nitrate (KNO_3), the crystallized salt.

(1) By boiling "Chilli saltpetre" ($NaNO_3$) with potassium chloride (KCl), $NaNO_3 + KCl = KNO_3 + NaCl$. The sodium chloride first crystallizes out, and then the long six-sided prisms of saltpetre. The common salt crystallizes in cubes. It is found native in some caves of India, especially in Bengal and Oude. In the above reaction it would be well to bear in mind that 100 parts of boiling water dissolves 200 parts of nitrate of potash, but only thirty-seven of chloride of sodium and when cold only thirty-six parts.

(2) It can be formed artificially by the oxidation of nitrogenous matter in the presence of very strong bases, like lime, as follows: large heaps of organic matter are moistened with stable washings, add plastering and the like. In wet countries it is protected from the rain, though freely exposed to the air, and finally the calcium nitrate is dissolved, when potassium nitrate is formed, $Ca(NO_3)_2 + K_2CO_3 = 2(KNO_3) + CaCO_3$.

50. Potassium nitrite (KNO_2) — a strong solution, one part of the salt to two of water, $KHO + HNO_2 = KNO_2 + H_2O$.

(1) By the action of nitrous acid on metallic oxides or hydrates, $KHO + HNO_2 = KNO_2 + H_2O$.

(2) By the action of heat on certain nitrates. Heat until a portion removed has a strong alkaline reaction; pour on a dry stone slab, and preserve in well stoppered bottles, $KNO_3 = KNO_2 + \overline{O}$; if the heat be carried too far, the KNO_2 breaks up into $K_2O + NO$.

(3) The nitrites of the alkalies can be made from nitrates by stirring their boiling solutions with a rod of zinc or cadmium (Schœnbein).

(4) By forming nitrous acid and passing it into KHO until it is completely saturated. The nitrous acid may be made by taking two parts of starch, eight parts of HNO_3, commercial sp. gr. 1.4, and eight parts of water and heating; as soon as the action begins, the flame is taken away. The fumes (N_2O_3) are passed first into an empty flask, and then into KHO. The The above will saturate five parts of KHO, sp. gr. 1.27. Or the nitrous acid can be made by gently heating HNO_3, sp. gr. 1.35, with an equal weight of arsenious acid, passing the gas through a U tube, cooled by cold water, to condense undecomposed nitric acid, and through a similar tube containing chloride of calcium, H_2O, $N_2O_5 + As_2O_3 = H_2O + As_2O_5 + \overline{N_2O_3}$. It will be noticed, that if N_2O_3 is passed in, a nitrite is formed, but if from any reason NO_3, a nitrate will be formed; the truth is, a mixture of nitrate and nitrite is formed — *strong* alcohol dissolves the nitrite, while the nitrate is practically insoluble.

51. Potassium sulphocyanide (KCyS), one part of the salt in twenty of water. If we have this salt, the corresponding ammonium salt is not required; they are made alike, and one description answers for both.

(*1*) See prussic acid. By neutralizing HCNS with NH_4HO, or by digesting HCN with yellow ammonium sulphide.

(*2*) KCyS is formed by fusing KCy and S; or by fusing $3(K_4FeCy_6) + K_2CO_3 + 2S$ in a covered crucible; the mass dissolved and the salt crystallized form the solution, as follows: $K_4FeCy_6 + 6S = 4KCyS + Fe (CyS)_2$.

(*3*) A mixture of 750 parts of NH_4HO and 100 parts of CS_2 and 750 parts of C_2H_6O (86 %) is distilled down one-half; the residue is evaporated to the point of crystallizing, as follows: $4NH_4HO + CS_2 = NH_4CNS + (NH_4)_2S + 4H_2O$.

52. Potassium hydrogen sulphate ($KHSO_4$). (See problems.)

53. Potassium sulphate (K_2SO_4). (See problems.) Take one part of the salt for 200 of water; for a strong solution, one to twelve.

54. Platanic chloride ($PtCl_4$). See gold chloride, No. 34.

55. Sodium acetate ($NaC_2H_3O_2$), one part of the salt to five of water. Made by adding acetic acid to a concentrated solution of Na_2CO_3 until all effervescence ceases, $2(HC_2H_3O_2) + Na_2CO_3 = 2Na(C_2H_3O_2) + H_2CO_3$.

56. Sodium carbonate (Na_2CO_3), the dry salt and a solution containing one part of the salt in five parts of water (Leblanc process).

(*1*) Common salt and sulphuric acid are heated in a furnace —"salt-cake" (Na_2SO_4) is formed, $2(NaCl) + H_2SO_4 = Na_2SO_4 + 2HCl$.

(*2*) The salt cake thus formed is heated with limestone and coal, when two changes take place: (*a*) $Na_2SO_4 + C_2 = Na_2S + 2(CO_2)$, and $CaCO_3 + C = 2(CO) + CaO$. (*b*) The CaO reacts upon Na_2S in the presence of carbonic anhydride, $Na_2S + CaO + CO_2 = Na_2CO_3 +$

CaS, producing a *soluble* sodic carbonate and an *insoluble* calcic sulphide. The filtrate is purified and the waste products utilized.

57. Sodium biborate [$Na_2O(B_2O_3)_2$]. The salt is dried and pulverized. By fusing sodic carbonate with boracic acid, the latter acid expelling the carbonate anhydride from the sodic carbonate, the residue is then dissolved and crystallized. It originally came from India and Thibet, where it was obtained in crystals from the waters of certain lakes, under the native name of *tincal*. Much of the boracic acid comes from the volcanic districts of Northern Italy.

58. Sodium hydrate (NaHO), one part of solid to nine of water.

Potassic hydrate (KHO) is prepared like the sodium hydrate.
(*1*) By burning potassium in pure dry oxygen, and treating the oxides formed, with water.
(*2*) By igniting a mixture of nitre (one part) and copper (three parts) in a copper vessel, and dissolving the residue in water.
(*3*) The commercial method — by the action of calcic hydrate [$Ca(OH)_2$] on a dilute solution of potassic carbonate (pearl ash K_2CO_3), $K_2CO_3 + Ca(OH)_2 = 2(KHO) + CaCO_3$. The solution of potassic hydrate is decanted from the *insoluble* calcic carbonate, and evaporated down in a silver basin. It should not give a precipitate with *baryta water, ammonic oxalate, silver nitrate*, nor *ammonic sulphide*.

59. Sodium thiosulphate ($Na_2S_2O_3$) [hyposulphite], one part of the salt in forty of water.
(*1*) By treating a solution of sodic sulphide with sulphurous acid.
(*2*) By digesting together sulphur and sodic sulphide.

(*3*) By exposing the calcic sulphide of the gas or alkali works to the air, when a calcic hyposulphite is formed by oxidation; if this is treated with sodic carbonate, sodic hyposulphite is formed. A stronger acid will not liberate free acid, but it is decomposed in sulphur and sulphurous acid, $H_2S_2O_3 = S + H_2SO_3$. (*a*) This reaction distinguishes it from sulphurous acid. (*b*) It dissolves silver chloride, forming silver sodic thiosulphite ($NaAgS_2O_3$). (*c*) A salt of ruthenium, rendered alkaline with ammonia, turns thiosulphuric acid a deep red color.

60. Sodium hypochlorite ($NaClO$). Agitate one part of good bleaching powder with ten parts of water. Add a solution of sodium carbonate, as long as a precipitate is formed; allow the solid matter to subside and siphon off.

61. Di-Sodic phosphate (Na_2HPO_4), one part of the salt in ten of water, and also crystals. By adding sodic carbonate to the tetrahydric calcic phosphate, obtained during the process of making phosphorous, $H_4Ca2(PO_4) + 2(Na_2CO_3) = CaCO_3 + H_2O + CO_2 + (Na_2HPO_4)_2$.

62. Sodium phosphomolybdate. (See alkaloids.) Use Sonnenschein's reagent for acid solutions of alkaloids. The yellow precipitate formed on mixing acid solutions of ammonium molybdate and sodium phosphate (the ammonium phosphomolybdate) is well washed, suspended in water, and heated with sodium carbonate until completely dissolved. The solution is evaporated to dryness, and the residue gently ignited till all the ammonia is expelled, sodium being substituted for ammonia. If blackening occurs from the reduction of molybdenum, the residue is moistened with nitric acid and heated again. It is now dissolved in water and nitric acid added to strong acidulation the solution being made ten parts to one part of the residue. Keep out of contact with vapor of ammonia, both during its preparation and when preserved for use.

63. Sodium sulphide (Na_2S), one part of the solution of soda, saturated with hydrosulphuric acid, to one part of unchanged soda solution.

64. Sodium sulphite (Na_2SO_3), one part of the salt to five parts of water. Made by passing SO_2 over crystals of Na_2CO_3.

65. Silver nitrate ($AgNO_3$), one part of the salt in thirty of water. Take a silver coin, dissolve it in nitric acid; you have the nitrates of copper and silver. Precipitate the silver nitrate by hydrochloric acid, as a silver chloride ($AgCl$), while the copper remains in solution and can be poured off; wash by decantation until the water gives no traces of copper with ferrocyanide or with ammonia. Reduce the silver chloride to *pure* metallicsilver by heating with Na_2CO_3 on charcoal, not taking too large a quantity at a time; or by means of zinc and hydrochloric acid, the nascent hydrogen combining with the chlorine of the AgClto form HCl, leaving the silver in a black, finely divided state. Dissolve the silver in the smallest possible quantity of strong nitric acid; evaporate to dryness. If you take *strong* nitric acid and dissolve the silver coin in it, the silver nitrate is insoluble in the cold and separates out; the process is rapid, but wasteful.

66. Stannous chloride ($SnCl_2$), one part of the crystallized salt in six parts of water, acidulated with hydrochloric acid. Boil granulated tin in concentrated HCl in a flask, keeping the tin always in excess, until H ceases to be evolved; then dilute the solution with four times its volume of water, containing a little hydrochloric acid, and filter. Keep the filtrate in a well stoppered bottle containing some fragments of tin.

67. Solution of indigo. Take one part of finely pulverized indigo, five parts of fuming H_2SO_4, added slowly and small portions at a time, keeping the mixture well stirred. Let it stand forty-eight hours, and pour it into twenty or thirty times its volume of water, filter, and keep the filtrate for use.

68. Zinc (Zn), granulated metal. Melt Zn in an iron ladle, and pour in a small stream from a height of six or eight feet into a jar of cold water; dry.

The reagents for the alkaloids are described where they occur.

It is a very valuable exercise for the student to make the reagents, as a laboratory practice, *and always to test their purity.* In all his work, he must bear this thought in mind, "Trifles make perfection, but perfection is no trifle."

CHAPTER II.

DEFINITIONS.

Chemistry is the science that describes the properties of the different kinds of matter, and teaches the laws which regulate their union and separation.

An atom is the smallest particle of matter capable of entering into or existing in a state of chemical combination.

A molecule is the smallest particle of matter capable of existing in a free state.

An element is a body that yields but one kind of matter; e. g., oxygen, iron, nitrogen. The elements now known are about seventy in number, and are divided into three classes — metals, non metals, and semi metals.

The metals are electro positive, form with oxygen basic anhydrides, and are good conductors of heat and electricity.

The non-metals are electro negative, form with oxygen acid anhydrides, and are non-conductors of heat and electricity.

The semi-metals resemble the metals in their physical properties, and the non-metals in their chemical properties.

Oxides that do not contain hydrogen are called *anhydrides.* The metallic oxides, like (K_2O) potassium oxide, (CaO) calcium oxide, (HgO) mercuric oxide, are called *basic anhydrides;* and the non metallic oxides, like (SO_3) sulphuric anhydride, (N_2O_5) nitric anhydride, are called *acid anhydrides.*

A radical is the residue of a molecule, and acts like the atom, as a unit (HO) hydroxyl.

Synthesis is the process of uniting two or more bodies to form a chemical compound.

Analysis is the process of separating a compound into its elements.

Chemical changes in analytical operations occur in the following ways:

1. Combination or synthesis; as, $C + O_2 = CO_2$.
2. Dissociation or analysis; as, $HgO + heat = Hg + \overline{O}$.
3. Transposition or metathesis; as, $HgCl_2 + (KI)_2 = HgI_2 + (KCl)_2$.
4. Oxidation or reduction.

In this book the sign $\overline{}$ placed over a symbol indicates that a gas is produced or evolved; the sign $\underline{}$ placed under a symbol that a solid is formed and precipitates.

An acid is a compound of hydrogen with a negative atom or radical; as, (HCl) hydrochloric acid, (H_2O, SO_3) sulphuric acid.

Acids are electro-negative binaries, which are generally formed by the union of oxygen with non metal.

A binary compound is the union of two elements. Such compounds are named by affixing the termination *ide* to the nonmetallic, or electro-negative element, and prefixing the name of the metal, or the electro-positive element; as, mercuric iodide (HgI_2).

Bases are electro-positive binaries which are formed by the union of oxygen with a metal; as, K_2O, CaO.

If four acids of an element exist, the prefix *hypo* is given to the lowest; as, hypochlorous acid (HClO), *ous* the next higher; as, chlorous acid ($HClO_2$), *ic* the next; as, chloric acid ($HClO_3$), *per* the highest; as, perchloric acid ($HClO_4$).

From the *ic* acid *ate* salts are formed; if we replace the hydrogen of chloric acid above given by potassium, we have ($KClO_3$) potassium chlor*ate*. From the *ous* acid *ite* salts are formed. Chlorous acid ($HClO_2$) replacing the H by K, we have potassium chlor*ite* ($KClO_2$).

Monobasic acid contains but *one* atom of replaceable hydrogen, and therefore can form only *one* series of salts; as, nitric acid (HNO_3).

Dibasic acid contains *two* atoms of replaceable hydrogen, and can form *two* series of salts (normal and acid salts), sulphuric acid (H_2SO_4), normal sulphate of potassium (K_2SO_4), acid sulphate of potassium, or bisulphate ($KHSO_4$).

Tribasic acid contains *three* atoms of replaceable hydrogen and can form *three* series of salts (normal salts and two series of acid salts), (H_3PO_4) phosphoric acid.

Normal salt is one in which the whole of the replaceable hydrogen has been displaced by a metal (K_2SO_4).

Acid salt is one in which only part of the replaceable hydrogen has been displaced by a metal ($KHSO_4$).

Double salt is one in which the hydrogen atom is displaced by atoms of two different metals; as, common alum Al_2O_3, $3(SO_3)$ $+ K_2O$, SO_3, or $Al_2K_2S_4O_{16}$, or $AlK (SO_4)_2$. *Notice* that it takes as many molecules of anhydrous acid to combine with the metal as there are of the oxide. In the above we have $Al_2 O_3$ or three oxygen, and also $3(SO_3)$; in K_2O, one oxygen and one SO_3.

A basic salt is a combination of a salt with a basic oxide: $3PbO$, $N_2O_5 = Pb(NO_3)_2$.　$2(PbO) = Pb_3(NO_3)_2$, O_2.

A salt is formed by dissolving a metal or its oxide in an acid. *Note*, an acid never combines with a metal, but with the oxide of the metal; as, $ZnO + SO_3 = ZnSO_4$. Or a *salt* is formed by the substitution of a metal for the hydrogen in an acid; as, H_2O, SO_3; if we replace the H_2 by Zn, we have a salt of zinc, ZnO, SO_3.

Symbols are letters used to represent the elements; as, (Cl) chlorine, (Br) bromine.

Formula represents a molecule of a substance by its symbols; as, (HCl) hydrochloric acid.

A reaction is representing chemical changes by their symbols:

Zinc + sulphuric acid = zinc sulphate + hydrogen.

$$Zn + H_2SO_4 = ZnSO_4 + \overline{H}_2.$$

TABLE OF ATOMIC WEIGHTS.

$H = 1.$

I.	II.	III.	IV.	V.	VI.	VII.	VIII.
Li=7.01	Be=9.09	B=10.91	C=11.97	N=14.02	O=15.96	Fl=18.98	
Na=23	Mg=23.96	Al = 27.01	Si=28.2	P=30.96	S = 31.98	Cl=35.37	
K=39.02	Ca=39.99	Sc=43.98	Ti=49.85	Vd=51.26	Cr=52.01	Mn=53.91	Fe = 55.91 Ni = 57.93 Co = 58.89
Cu=63.17	Zn=64.91	Ga=68.85	—	As=74.92	Se=78.8	Br=79.77	
Rb=85.25	Sr=87.37	Yt=89.82	Zr=89.37	Nb=94.	Mo=95.53	—,	Rh=104.06 Ru=104.22 Pd=105.74
Ag=107.68	Cd=111.77	In=113.4	Sn=117.7	Sb=119.96	Te=127.96	I=126.56	
Cs=132.6	Ba=136.76	La=138.53	Ce=140.42	Di=146.18	Tb=148.8	—	
—	—	—	—	Er=165.89	—		—
—	—	Yb=172.76	—	Ta=182.14	W=183.61	—	Ir =192.65 Pt =194.42 Os =198.49
Au=196.16	Hg=199.71	Tl=203.72	Pb=206.47	Bi=207.52	Ng=214	—	
—	—	—	Th=233.41	—	Ur=238.48	—	

Atomic weight of an element is the weight of one atom of an element compared with the weight of an atom of hydrogen. (See table of Atomic Weights.)

Molecular weight is the sum of the weights of all the atoms comprising the molecule of a substance. $[(H_2O) — H_2 = 2, O = 16; 2 + 16 = 18.]$

Valency or atomicity are terms used to express the combining power of one atom of an element as compared with that of an atom of hydrogen, and is indicated by one or more dashes attached to the symbols — Cl' = univalent, O'' = divalent, N''' = trivalent, C^{IV}= quadrivalent. Those elements whose valency can be represented by 1, 3, 5, 7, are called *perissads;* those whose valency is 2, 4, 6 or 8, are called *artiads.* The sum of the bonds in a stable, saturated molecule, is always an even number; as, H_2O or H-O-H, 2 hydrogen and 2 oxygen, making 4 in all.

CHAPTER III.

TESTS IN THE DRY WAY.

The following hints will be of use to the student:
1. Have everything neat and clean.
2. Never put anything away dirty.
3. Know what you do and why you do it.
4. Do the work *yourself*.
5. Keep accurate notes at the time.
6. Take the smallest possible quantity of the assay.
7. Note the characteristic tests, and remember two of them.
8. Look up the properties and reactions of the substance in some work of reference.
9. Never quit a substance until you have mastered it.
10. It is not *how much* you do, but *how well*.

NOTICE.—Do not expose yourself needlessly to the vapors of the laboratory—as, chlorine, hydrogen sulphide, arsenic, etc. The bad effects may not be perceptible immediately.

The following is the most convenient way to keep a note book:

NAME.	SYMBOL.	STATE.	REACTION.	REMARKS.

The above occupying two pages of a note book.

Under *name* put the name of the substance.

Under *symbol* put the symbol of the substance.

Under *state* describe its physical appearance, etc.

Under *reaction* express by an equation the chemical change that has taken place by heating it, etc.

Under *remarks* put how it is manufactured, etc., also, special peculiarities ; as,

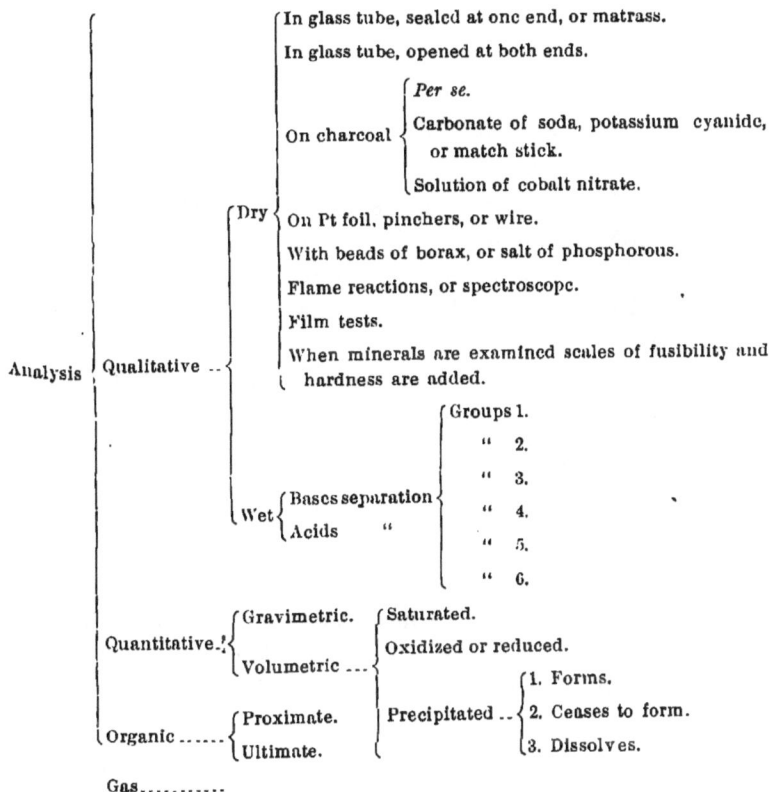

Analysis
- Qualitative
  - Dry
    - In glass tube, sealed at one end, or matrass.
    - In glass tube, opened at both ends.
    - On charcoal
      - *Per se.*
      - Carbonate of soda, potassium cyanide, or match stick.
      - Solution of cobalt nitrate.
    - On Pt foil, pinchers, or wire.
    - With beads of borax, or salt of phosphorous.
    - Flame reactions, or spectroscope.
    - Film tests.
    - When minerals are examined scales of fusibility and hardness are added.
  - Wet
    - Bases separation
      - Groups 1.
      - " 2.
      - " 3.
      - " 4.
      - " 5.
      - " 6.
    - Acids "
- Quantitative
  - Gravimetric.
  - Volumetric
    - Saturated.
    - Oxidized or reduced.
    - Precipitated
      - 1. Forms.
      - 2. Ceases to form.
      - 3. Dissolves.
- Organic
  - Proximate.
  - Ultimate.
- Gas............

I. Note the physical properties of the substance to be tested.

(a) Its state, solid or liquid.

(b) Its form, crystalline or amorphous.

(c) Its color, lustre, hardness, etc.

(d) If it suffers any change when exposed to the air.

II. Heat a portion of the substance, finely pulverized, in a dry glass tube, closed at one end.

I. The substance suffers no change:

Absence of volatile bodies, water, organic compounds, and of those which change color on heating.

II. The substance changes color:

Cu and *Co* salts blacken at high heat. *Organic* bodies blacken from separation of carbon. SnO_2, brown, while hot; yellow, when cold. Fe_2O_3 and salts, red to black, while hot; brown, when cold. ZnO and salts, yellow, while hot; white, when cold. Bi_2O_3 and salts, orange to red brown, while hot; pale-yellow, when cold. PbO and salts, yellow, while hot, fusible at a very high temperature.

III. The substance sublimes:

H_2O condensing in the cold part of the tube. *Hg*, gray-metallic, if heated with Na_2CO_3. $HgCl_2$ melts and forms white crystalline sublimate. Hg_2Cl_2, yellow, while hot; white, when cold. *HgS* black sublimate, red on rubbing. *As*, steel gray; garlic odor. *I*, violet vapor, blue-black sublimate. NH_4 salts, white, heat with soda—lime, $\overline{NH_3}$. *S*, reddish drops. *Sb*, yellow mass — H_2S gives orange. As_2O_3 white-octahedral crystals (under microscope). As_2S_3, black, while hot; reddish yellow, when cold. *Alums* and borax puff up.

IV. The substance fuses:

Most *alkali salts*. Many salts dissolve in their water of crystallization, when heated, becoming solid again by evaporation.

V. Substance evolves a gas or vapor:

O indicates the presence of nitrate, chlorate, bromate, iodide, peroxide — ignites glowing coal. H_2S from sulphides, some sulphides blacken lead paper, odor of rotten eggs. SO_2, recognized by its odor and bleaching action from sulphides and some sulphates. NH_3 from compounds which easily decompose, odor and by litmus. Cy or CN, odor of peach kernels. *Nitrogen oxides* from nitrates and nitrites, reddish-brown, acrid vapors.

III. In a glass tube open at both ends.

S and sulphides yield SO_2, odor and by litmus paper.
As yields a sublimate; examine with glass.
Hg, metallic Hg; sublimate.
Bi, dark brown, while hot : lemon yellow, when cold.
Sb, white sublimate, Sb_2O_3 and Sb_2O_5.
$\{Se\}$ are best detected by Bunsen's film test with H_2SO_4 (red
$\{Te\}$ Te) (green Se).*

To detect *S* or *P*, heat the well dried substance in a tube of *thin* glass with Na or Mg, and place on silver coin, moistened with a drop of water. *S* gives black hepar.† *P* gives H_3P, odor.

IV. Heated in the blowpipe flame on charcoal.

1. The substance decrepitates:

Any crystals containing H_2O — as, NaCl, if finely pulverized, the decrepitation is avoided.

2. The substance deflagrates:

Nitrates, chlorates, permanganates, iodates, hypophosphites, etc.

---

*Burn the substance on asbestos under a test tube that contains about a half inch of water. Have this test tube fit inside of another that contains a few drops of the *strongest* sulphuric acid ($H_2O_2(SO_3)$), when the above reaction will be given, by forming the oxide on the smaller tube, and having this dissolved by the sulphuric acid by inserting it into the larger tube.

† It is moistened sometimes with hydrochloric acid to bring out the sulphide stain: always use the dilute acid, about one to twelve of water, as the strong acid itself blackens the silver coin by leaving it on the coin for a few minutes.

3. The substance fuses, and is absorbed by the charcoal:
Salts of the alkalies, and some of alkaline earths.

V. The substance is moistened with a solution of cobalt nitrate, and again strongly ignited.

This test can be used only for substances that are nearly or quite white color after igniting in the O. F. The quantity of the solution depends upon its concentration; dilute solutions yield the best results, three or four drops will be enough.

BE SURE AND NOT BE DECEIVED BY THE DECOMPOSITION OF THE COBALT SOLUTION WHEN HEATED.

It is good for Al, Zn, Mg, Sn; the Al and Mg reactions are prevented by the presence of colored metallic oxide, which generally produce a gray or black mass, unless present in very small quantities.

I. The mass is colored:
ZnO, yellowish green.
SnO, bluish green.
SrO, CaO, gray.
BaO, brick red.
MgO, flesh color or pink.
Al_2O_3, SiO_2 and phosphates, blue.
Sb_2O_5, dirty dark green.

2. The substance is infusible and phosphorescent:
Al_2O_3, MgO, ZnO (yellow while hot), not alkaline, Ba, Sr, Ca, Mg, alkaline to test paper.

3. The substance forms an incrustation on charcoal.
Pb, lemon-yellow, while hot; if in thin layers, it may be bluish white, burns with bluish flame.
Zn, yellow, while hot; white, when cold, greenish flame.
Bi, orange yellow, while hot; lemon yellow, when cold.

Sn, faint yellow, while hot; white, when cold.

Cd, red brown; dark yellow flame.

As, white; blue flame.

Sb, white; pale green flame.

VI. Heated with Na₂CO₃ on charcoal in reducing flame.

Pb, Cu, Sn, Au, malleable bead.

Bi, Sb, brittle bead.

Fe, Ni, Co magnetic grains.

VII. Flame reactions.

The chlorides being more volatile than other forms, therefore heat a portion of the substance, moistened with HCl, on a piece of Pt wire in the outer flame of a Bunsen burner, or in the oxidizing flame of a blowpipe.

Na, reddish yellow.

K, violet.

Ca, brick red to yellowish red.

Sr, crimson.

Li, carmine red.

Ba, green.

Cu, emerald green.

B_2O_3, yellowish green, when moistened with H_2SO_4.

Zn, whitish green.

As, light blue.

$CuCl_2$, Pb, Se, azure blue.

The use of cobalt glass to shut off the sodium rays will do the student more harm than good; when very *small* quantities are at hand, the spectroscope may be used.

VIII. Treatment of the substance with borax and microcosmic salt.

A small platinum wire is fused into a glass tube or rod, and the end heated, and dipped into the borax or salt of phosphorus, and heated to a colorless bead. The substance to be tested is touched by the hot bead and a portion taken up, and again heated in oxidizing and reducing flames. The chemistry of the process is quite simple. The metallic oxides are dissolved in the bead, and forms a colored glass. Of course, the color will vary with the amount of the substance taken. Certain bodies, as the alkaline earths, dissolve in borax, forming beads, which, up to a certain period of saturation, are clear. When these beads are brought into the reducing flame, and an intermittent blast is used, they become opaque. This is what is known as *flaming*. Tin, lead and silver are the reducing agents used in the beads. These metals cannot be used upon Pt wire, as they would form an alloy with the Pt. The beads are formed in a loop of Pt wire, and while hot shaken off into a porcelain plate, and when a sufficient number is obtained, they are fused on charcoal into a large bead, which is charged with the substance to be tested, and then with the tin or other metal. The great use of the beads, therefore, is to form a flux for the oxide of the metals.

Fluxes are used in the following cases :

1. To cause the fusion of a body, either difficultly fusible or infusible by itself.

2. To fuse foreign substances mixed with a metal, in order to allow the latter to separate by its difference of specific gravity.

3. To destroy a compound into which an oxide enters, and which prevents the oxide being reduced by charcoal.

4. To prevent the formation of alloys of some metals with others.

5. To scorify some of the metals contained in the substance to be assayed, and to obtain the others alloyed with a metal contained in the flux.

6. To obtain a single button of the metal, which otherwise would be obtained in globule.

COLOR OF A BEAD OF BORAX OR MICROCOSMIC SALT.

COLOR OF TILE BEADS.	WITH PHOSPHOROUS SALT.		WITH BORAX.	
	OXIDIZING FLAME.	REDUCING FLAME.	OXIDIZING FLAME.	REDUCING FLAME.
Brown to yellow.	Fe, Ni, U, Ag.	Fe, TiO_2.	Fe, U, Pb, Bi, Sb.	TiO_2, Mo.
Red.	Fe, Ni, Cr, when hot and strongly saturated.	Fe, TiO_2. W, containing Fe (blood red) when hot.	Fe, Ni, when cold (reddish brown).	Cu (opaque), strongly saturated.
(Amethyst.) Violet.	MnO_2, D_2O_3, hot and cold.	Tl, Nb, cold, strongly saturated.	Mn, D_2O_3, Ni + Co, hot and cold.	
Blue.	Co, Cu, hot and cold.	Co, W, Nb, hot and cold, strongly saturated.	Co, hot and cold. Cu, cold.	Co, hot and cold.
Green.	Cu, Mo, Fe + Co or Cu, hot. Cr, U, cold.	Cr, U, Mo, cold.	Cr, cold. Cu, Fe + Co or Cu, hot.	Fe, U, Cr, hot and cold.
Gray.		Ag, Zn, Cd, Pb, Bi, Sb, Tc, Ni, cold.		As with phosphorous salt.

NOTE.—Reduction takes place more easily with phosphorous salt. In general, the behavior of the various bodies is quite similar with borax and phosphorus. Salt of phosphorus is especially good for the detection of silica.

FILM TESTS.

Name.	Oxide incrustation with AgNO₃ and NH₃	Iodide incrustation and coating.	Iodide incrustation with NH₃ (blown upon it).	Sulphide incrustation and coating.
Antimony.	Black insoluble in NH₃.	Orange, breath'd upon, disappears for a time.	Disappears permanently.	Orange.
Arsenic.	Lemon yellow, or brownish red; soluble in NH₃.	Orange yellow, breath'd upon, disappears for a time.	Disappears permanently.	Lemon yellow.
Bismuth.	White.	Bluish brown, with light red coating; breath'd upon, disappears for a time.	Rose red to orange yellow; chestnut bro'n when dry.	Umb'r br'n with coffee, bro'n coating.
Mercury.		Carmine and lemon yellow; breath'd upon, does not disappear.	Disappears for a time.	Black.
Lead.	White.	Orange yellow to lemon yellow; breath'd upon, does not disappear.	Disappears for a time.	Through brownish red to black.
Cadmium.	Coating becomes blue black.	White.	White.	Lemon yellow.

Take an evaporating dish, half fill it with water, the object of which is to keep it cold; now burn a small piece of the substance on an asbestos fibre under the dish; if the fibre is held in the reducing flame, there will be a *metallic incrustation;* if in the upper oxidizing flame, an *oxide incrustation* on the bottom of the dish. The dish is now quickly turned over to empty the water, care being taken not to get the bottom wet. This incrustation can be now treated with stannous chloride (a drop or two), then with caustic soda, silver nitrate, and blowing ammonia upon it. The *iodide incrustation* is formed by placing the oxide incrustation on the dish, and burning under the dish a piece of asbestos, moistened with iodine dissolved in alcohol. *Sulphide incrustation* is obtained from the last by blowing ammonic sulphide upon it, and then gently warming the disk or capsule.

The metallic incrustation and coating, the oxide incrustation and coating, the oxide incrustation with SnCl₂ and, also, with SnCl₂ and NaHO, are omitted, as not being characteristic. The above tests are very delicate, and a very nice way to identify As from Sb. The arsenic reactions of the above are given by a piece of wall paper that contains it, of the size of half an in inch by one-fourth of an inch. A casserole is more convenient than an evaporating dish in all this work.

The following is the best method for commencing an analysis in the dry way:

1. Test the solution by flame reaction on platinum wire.

2. Heat on platinum foil, volatile and odorous bodies are identified; bodies rich in oxygen deflagrate, those containing water decrepitate.

3. Heat on charcoal. The earths and alkaline earths glow and are infusible. Some change color when hot and cold, give a characteristic odor.

4. Heated on charcoal, moistened with a dilute solution of cobalt, nitrates change their color.

5. Heated on charcoal with sodium carbonate or potassium cyanide, or both, in the reducing flame; metals may or may not be magnetic.

7. In contact with a salt of phosphorus, or borax bead, gives colored reactions in oxidizing and reducing flames.

8. In contact with zinc and dilute sulphuric or hydrochloric acid, gives a colored fluid with some of the acids.

9. If an alloy, dissolve it in dilute nitric acid by means of heat. Antimony and tin are converted into oxides (insoluble in an excess of the acid). All the normal nitrates are soluble in water.

10. In solutions. (a) Test with litmus paper, neutral acid, or alkaline. (b) Evaporate a portion to dryness on platinum foil or evaporating dish. A *neutral* solution, as a rule, contains salts of the *alkalies* or *alkaline earths*, while salts of nearly all the *metallic oxides* have an *acid* reaction. Metals are now carefully separated into groups, and the groups separated according to the scheme on page 23, when the acids are tested. The presence or absence of certain bases will often give a clue to the acids; as, for instance, if you have a solution of lead or the alkaline earths, you need not look for sulphuric acid.

TESTS IN THE WET WAY.

In qualitative chemical analysis, the metals or bases are commonly divided into five groups.

1. Those metals forming insoluble chlorides in water are precipitated from their solutions (better if nitrates) by the group reagent, *hydrochloric acid.* Pb, Ag, Hg, and the rare elements, tungsten (as tungstic acid), venadium and thallium (as chloide). This is called the *silver group.*

It is generally omitted when Ag is not present. as Pb and Hg are as easily separated by the next group.

SEPARATION OF THE SILVER GROUP.

LEAD.	SILVER.	MERCURY.

Add hydrochloric acid.

PRECIPITATES

Lead chloride.	Silver chloride.	Mercurous chloride.

Boil with water.

SOLUTION.	RESIDUE.	
$PbCl_2$, on cooling, acicular crystals. Add H_2SO_4, $\underline{PbSO_4}$. Reduce with carbon and Na_2CO_3 to malleable Pb; dissolve in dilute HNO_3 and add K_2CrO_4, $\underline{PbCrO_4}$, yellow.	AgCl.	HgCl.
	Add NH_4HO.	

SOLUTION.	RESIDUE.
AgCl, add HNO_3 AgCl. Reduce on charcoal to metallic Ag; dissolve in dilute HNO_3, add K_2CrO_4, $\underline{Ag_2CrO_4}$, red.	NH_2Hg_2Cl, black.*

*NOTE.—Dry, and put in a hard glass tube and heat it with carbon and Na_2CO_3, sublimate of Hg. Dissolve this in dilute HNO_3, add KI, greenish $\underline{HgI}$; but if the Hg is dissolved in strong HNO_3, you get a red $\underline{HgI_2}$.

Lead—Pb^{II-IV}, 207: TESTS.†

(1) Gives yellow coating on charcoal before the blowpipe.

(2) Insoluble in strong HCl and H_2SO_4, but dissolves in HNO_3, $3Pb + 8HNO_3 = 3Pb(NO_3)_2 + \overline{2NO} + 4H_2O$.

† The atomic weights are uncertain, as yet, and fractions are omitted.

-5

(3) Zn precipitates it from its neutral solution as metallic
. lead.

(4) $K_2Cr_2O_7$ precipitates yellow $PbCrO_4$.*

(5) $(NH_4)_2S$ precipitates black PbS.

(6) H_2S precipitates black PbS.

(7) H_2SO_4 precipitates white $PbSO_4$, insoluble in acids.

(8) HCl precipitates with $PbCl_2$, soluble in boiling H_2O.

NOTICE — While dilute HNO_3 is the best solvent for Pb, Pb_3O_4
or "red lead" or minium is not soluble in HNO_3, but there is
formed $Pb(NO_3)_2$, and PbO_2 is left undissolved. No. (6) is the
most delicate test.

Silver—Ag^I, 108:   TESTS.

(1) Easily reduced to metallic Ag before the blowpipe.

(2) HNO_3 is the best solvent.

(3) HCl precipitates white AgCl. insoluble in HNO_3 but
soluble in NH_4HO.

(4) $K_2Cr_2O_7$ precipitates red Ag_2CrO_4.

(5) Zn, Cu, Fe, Mg, Pb, Hg, etc., reduce it to the metallic
state.

(6) H_2S and $(NH_4)_2S$ precipitate black Ag_2S.

---

*The student is expected to complete all the reactions in his note book, as follows:
If the chloride of lead ($PbCl_2$) be taken, $2PbCl_2 + K_2Cr_2O_7 + H_2O = 2(PbCrO_4) + 2(KCl) + 2(HCl)$. If the nitrate ($Pb(NO_3)_2$) be taken, $2Pb(NO_3)_2 + K_2Cr_2O_7 + H_2O = 2(PbCrO_4) + 2(KNO_3) + 2HNO_3$). If any acid, as HNO_3, be taken, there is *one* H replaceable: now we can put in place of that H any *monad* element—as, $K = KNO_3$. but if we put in a dyad, as Pb, we must double the acid radical, as $Pb(NO_3)_2$; by bearing this rule in mind, we can write the salt of any acid or element. If we want to write a sulphate of the ferric salt of iron, for instanae, we know the ferric salt is a triad. To find how many times to take the acid radical, multiply the valency of the element by the number of times it is taken, and divide this product by the replaceable H of the acid. In the above, Fe being a triad and being taken twice. it would have six bonds to satisfy. Now, H_2SO_4 has two replaceable H, and you must take it three times to satisfy these bonds. The note about the number of times the anhydrous acid is taken to combine with the metallic oxide must be remembered here. See page 20. When an element has two valencies, the *ic* salt is the higher.

NOTICE — The (3) is the most delicate. The nitrates, acetates and sulphates form permanent anhydrous crystals. The orthophosphate and arsenite, yellow; the arseniate, reddish brown; the iodide, yellow; the bromide, yellowish white; sulphide, black. The most of the salts are colorless.

Mercury — Hg^I, 200 : TESTS.

(*1*) Reduced in tube heated with carbon and Na_2CO_3.

(*2*) HNO_3, the best solvent.

(*3*) SO_2 precipitates gray metallic Hg.

(*4*) H_2S and $(NH_4)_2S$ precipitate black Hg_2S.

(*5*) HCl precipitates white Hg_2Cl_2 (calomel).

(*6*) KI precipitates greenish yellow HgI.

(*7*) NH_4HO precipitates black amido-compound, NH_2Hg_2 NO_3.

Notice *ous* salts of mercury readily change to the *ic*. If Ag is not present, this separation can be omitted.

The following facts must be considered in this group :

(*1*) HCl may expel $\overline{\overline{CO}}_4$, $\overline{H_2S}$, $\overline{HCN}$, which you must not fail to test.

(*2*) In the presence of hot or concentrated $Hg(NO_3)_2$, AgCl is not immediately precipitated.

(*3*) HCl may precipitate $BiCl_3$ or $SbCl_3$, because they are decomposed by water and break up into a soluble acid and an insoluble basic salt, $BiCl_3 + Bi_2O_3 =$ $3BiOCl$.

(*4*) HCl may precipitate boracic acid, inorganic and benzoic, uric organic acids. The boracic and benzoic are dissolved by hot water, the uric acid by HNO_3.

(*5*) The precipitate may be due to silicic acid dissolved in an alkaline solution.

This will show you the importance of examining your solution, and carefully notice its condition. In adding a reagent, it will be well to observe the following general rule :

Add a reagent until the last drop ceases to give a precipitate, and always test the filtrate to be sure it is all down.

2. The metals, which in a *slightly acid* solution form insoluble sulphides, are precipitated by the group reagent, *hydrosulphuric acid* (H₂S), Hg'', Pb, Bi, Cu, Cd, As, Sb, Sn, and the rare elements of the "Cu group," palladium, rhodium, osmium and ruthenium, in the "As group," gold platinum, iridium, molybdenium, tellurium and sellenium. This is called the *lead* and *arsenic group*.

SEPARATION OF THE LEAD AND ARSENIC GROUPS – FIRST METHOD.

Hg, Pb, Bi, Cu, Cd. | As, Sb, Sn. Pass H_2S.

PRECIPITATE.
HgS, PbS, Bi_2S_3, CuS, CdS, As_2S_3, Sb_2S_3, SnS. Boil with $(NH_4)_2S$.

RESIDUE.
HgS, PbS, Bi_2S_3, CuS, CdS. Boil with dilute HNO_3.

SOLUTION.
$(NH_4)_3AsS_3$, $(NH_4)_3SbS_3$, $(NH_4)_3SnS_3$. Add dilute HCl.

Residue branch (HgS, PbS, Bi₂S₃, CuS, CdS)

RESIDUE. HgS; test as before given for Hg.

SOLUTION. $Pb(No_3)_2$, $Bi(No_3)_3$, $Cu(No_3)_2$, $Cd(No_3)_2$. Evaporate to get rid of excess of acid; now dilute. Add H_2SO_4.

PRECIPITATE. $PbSO_4$; reduce to metallic lead on charcoal; test as before with KI or K_2CrO_4.

FILTRATE. $Bi_2(SO_4)_3$, $CuSO_4$, $CdSO_4$. Add NH_4HO in excess.

PRECIPITATE. $Bi(OH)_3$; dissolve in HCl, evaporate to a small bulk; add to a large volume of H_2O, $BiOCl$, white insoluble in H_2T.

FILTRATE. $Cu(OH)_2$, $Cd(OH)_2$; blue indicates Cu; add KCy until blue disappears. Add H_2S.

FILTRATE. $CuCy_2$.

PRECIPITATE. CdS yellow.

SECOND METHOD.
Add H_2S.
PRECIPITATE. CuS, CdS; add dilute H_2SO_4.

RESIDUE CuS. | **SOLUTION** $CdSO_4$. Add H_2S. CdS, yellow.

Solution branch (AsS₃, SbS₃, SnS₃)

PRECIPITATE. As_2S_3, Sb_2S_3, Sn_2S_3. Boil with HCl.

RESIDUE. As_2S_3, yellow. Boil with $KClO_3 + HCl$.

SOLUTION. H_3AsO_4. Add $NH_4HO + NH_4Cl + MgCl_2$. "Magnesia mixture." $MgNH_4AsO_4$, white.

SOLUTION. Sb_2Cl_6, $SnCl_4$. Dilute, place in an evaporating dish with Pt and Zn.

PRECIPITATES. Sb, Sn. Filter and wash to remove $ZnCl_2$. Now boil the precipitate with HCl.

RESIDUE. Sb; dissolve in $HNO_3 + HCl$; now add to $H_2O = SbOCl$, white. Dissolve in H_2T; add H_2S, (Sb_2S_3), orange red; reduce on charcoal to brittle bead, Sb.

SOLUTION. $SnCl_2$. Add $HgCl_2$, $2(HgCl)$ white. Reduce with $Na_2CO_3 + KCy$ on charcoal, ductile Sn.

SEPARATION OF THE LEAD AND ARSENIC GROUPS – SECOND METHOD.

Acidulate the solution with HCl as before, and pass H_2S through it. Heat it gently, and continue the operation for some time.

THE PRECIPITATE.

PbS, HgS, Bi_2S_3, CuS, CdS, As_2S_3, Sb_2S_3, SnS_2, SnS, brown; SnS_2, yellow. Wash the ppt. and boil with $(NH_4)_2S$.

RESIDUE.

PbS, HgS, Bi_2S_3, CuS, CdS. Boil with dilute HNO_3, dilute with water; add dilute H_2SO_4, until a precipitate ceases to be formed; when cold, add to the solution an equal bulk of methylated alcohol (CH_3OH); filter.

RESIDUE.

Contains Pb, Hg. Add $H_2C_4H_4O_6$, and NH_4HO in excess; boil and filter.

RESIDUE.	FILTRATE.
HgS. Test as before.	$PbSO_4$. Test by $K_2Cr_2O_7$. $PbCrO_4$, yellow.

FILTRATE.

Contains Bi, Cu, Cd. Boil to expel the alcohol; add NH_4HO in excess; boil and filter.

PRECIPITATES.	FILTRATE.
$Bi(OH)_3$. Dissolve in HCl and add to H_2O BiOCl, white.	Cu and Cd, blue, indicates Cu; add HCl in slight excess, pass H_2S; wash the ppt. with H_2S water; boil the ppt. with dilute H_2SO_4, and filter.

RESIDUE.	FILTRATE.
CuS; test as before.	Cd. Add NH_4HO and H_2S. CdS.

SOLUTION.

$(NH_4)_3AsS_3$, $(NH_4)_3SbS_3$, $(NH_4)_3SnS_3$.
1. Add HCl, precipitates SnS_2, Sb_2S_3, As_2S_3; filter and wash.
2. Digest the precipitate at a gentle heat with $(NH_4)_2CO_3$; filter.

RESIDUE.

SnS_2, Sb_2S_3; dissolve in boiling HCl; mix the acid solution with Zn in a Marsh's apparatus in order to generate H_3Sb.

Metallic mirror is form'd when a cold piece of porcelain is held in the ignited gas, indicates Sb. See page 43.

If there is any metallic ppt. on the Zn, detach it, and dissolve in HCl, dilute with H_2O; add $HgCl_2$, you may have a gray ppt. Hg or a white ppt. of Hg_2Cl_2; reduce to Sn on charcoal.

FILTRATE.

As. Add HCl, a yellow As_2S_3; confirm as before.

NOTE.—The common alcohol (ethylic C_2H_5OH) is used, adding ten per cent. of the methyl alcohol (CH_3OH). This is what is meant by methylated alcohol.

In this group it is well for the student to notice that H_2S is a *reducing agent:* that an excess of acid, especially free HNO_3, will decompose it; that As^V is not precipitated immediately by H_2S, and completely only from hot solutions. If the solution is very acid, it must be evaporated to expel the excess of acid. The normal condition of the solution is, that it should be slightly acid to litmus paper, made so by adding a few drops of HCl. It should be *acid* for two reasons: (1). To prevent the members of the 3d and 4th groups from being precipitated. (2) To insure the complete precipitation of this group.

All the bases precipitated by this group are more or less colored (yellow, orange, brown or black).

Mercury — Hg^{II}, 200 :   TESTS.

(1) Can be reduced in a tube as before described.

(2) $SnCl_2$, white, precipitates Hg_2Cl_2.

(3) H_2S or $(NH_4)_2S$ black precipitates HgS.

(4) $(KI)_2$, red, precipitates HgI_2, soluble in excess.

(5) Cu (metallic) precipitates Hg (metallic).

(6) NaHO or KHO, yellow, precipitates HgO.

NOTE — KI distinguishes between the *ic* and *ous* salts.

Lead — Pb^{II}, 207.   Same as on page 33.

Bismuth — Bi^{III}, 210 :   TESTS.

(1) Gives yellow coating on charcoal.

(2) $BiCl_3 + H_2O = BiOCl$, white — "pearl white."

(3) K_2CrO_4 precipitate syellow $Bi_2O(CrO_4)_2$ (not soluble in fixed alkali hydrates, differs from that of lead).

(4) H_2S or $(NH_4)_2S$ precipitates black Bi_2S_3.

(5) NaHO or NH_4HO precipitates white $Bi(OH)_3$.

NOTE — Bismuth has an unusual tendency to *basic* formations, the chlorides forming oxy-chlorides, etc. Nitric acid is the best solvent.

Copper — Cu^{II}, 63 :   TESTS.

(*1*) Flame, green to blue.

(*2*) Metallic Fe precipitates metallic Cu from solution.

(*3*) NaHO precipitates blue $Cu(OH)_2$; black on being boiled.

(*4*) NH_4HO precipitates and dissolves again $Cu(NO_3)_2 4NH_3$.

(*5*) H_2S and $(NH_4)_2S$ precipitates black CuS.

(*6*). $\begin{cases} K_3FeCy_6 \text{ precipitates yellowish green } Cu_3(FeCy_6)_2. \\ K_4FeCy_6 \text{ precipitates reddish brown } Cu_2FeCy_6. \end{cases}$

(*7*) $KC_2H_5COS_2 + H_2O$ (xanthate of potassium) precipitates brownish $Cu(C_2H_5COS_2)_2$, changes to a bright yellow — *the most delicate test.**

NOTE — Nitric acid the best solvent.

Cadmium — Cd^{II}, 112 :   TESTS.

(*1*) Before the blowpipe, brown incrustation.

(*2*) Metallic Zn precipitates metallic Cd.

(*3*) H_2S precipitates yellow CdS, insoluble in NH_4HO.

(*4*) $(NH_4)_2S$ precipitates yellow CdS.

(*5*) NaHO and NH_4HO precipitates white $Cd(OH)_2$.

(*6*) Na_2CO_3 precipitates white $CdCO_3$.

NOTE — Nitric acid is the best solvent.

On the following pages the *tests* and also a *comparison* of P, As, Sb, are given. It will be of value to students to *carefully* compare these tests for himself, and let him remember, that arsenetted hydrogen is exceedingly poisonous, and has no known antidote; therefore use only one-tenth of a grain of arsenic to test.

---

*Made by dissolving caustic potash in absolute alcohol, and adding bisulphide of carbon; filter, and dissolve the precipitate in water.

FREE EITHER NATIVE OR REDUCED.	REACTIONS: P—Atomic weight 31.
1. Heated in small tube (Berzelius).	Burns with white flame.
2. Heated with HNO_3.	With strong acid forms P_2O_5.
3. Heated with $NaClO$.	Melts, but suffers no change.
4. Heated with $HCl + KClO_3$.	Oxidizes to P_2O_3 or P_2O_5, depending on supply of O.
5. Heated with $(NH_4)_2S$.	Melts, but suffers no change.
6. Heated in open test tube.	Burns to P_2O_3 or P_2O_5, depending on supply of air. Free P yields luminous vapors with nascent hydrogen. See No. 12.

IN COMBINATION — DRY WAY.

7. Heated with C; $Na_2CO_3 + C$; with $Na_2CO_3 + KCy$.	Luminous in the dark.
8. Heated on charcoal.	Luminous in the dark.
9. On asbestos fibre.	....
10. Warm with $Cu + HCl$ (Reinsch's).	....
11. With $Zn + HCl$ {Marsh's. Davy's. (Na Hg).	....
12. Heat $NaHO$ or KHO or $CaHO$ } $+ P$.	H_3P { Self igniting in air, also characteristic odor. When passed into a solution of Cu, black Cu_3P_2; passed into $AgNO_3$, yields black precipitate of Ag. A very strong reducing agent.

IN NEUTRAL SOLUTIONS.

13. $AgNO_3 + NH_4HO$.	Orthophosphates, yellow Ag_3PO_4; others give white.
14. $CuSO_4 + NH_4HO$.	....
15. $MgCl_2 + NH_4Cl + NH_4HO$ (magnesium mixture).	$MgNH_4PO_4$.

IN ACID SOLUTIONS.

16. H_2S.	....
17. $(NH_4)_2Mo_3$.	Yellow ammonium phospho-molybdate.

6

REACTIONS: **As — Atomic weight, 75.**

*1......Sublimes at dull red heat to lustrous metallic coating.
2......Dissolves as As_2O_5, $3H_2O$.
3......Dissolves as Na_3AsO_4.
4......Dissolves as $AsCl_3$, and oxidizes to As_2O_3 and As_2O_5.
5......Slowly dissolves to $2(NH_4)_2SAs_2S_3$.
6......Oxidizes and sublimes to octahedral As_2O_3, soluble in water.

7......Is reduced to As, and sublimes.
8......Is reduced and burns with alliaceous odor.
9......Yields oxide film, becoming yellow with $AgNO_3 + NH_4HO$; iodide
      incrustation and coating; egg yellow; breathed upon, disappears
      transitorily; sulphide incrustation and coating, lemon yellow.
10......Is reduced to As, forming a steel gray film (Cu_5As_2) (Reinsch's).

11......Evolves H_3As
{
Burns with a livid flame, yields As_2O_3.
Deposits on porcelain plate lustrous mirror.
Mirror heated volatilizes rapidly.
Mirror is dissolved by chloride of lime.
Heated in open tube yields As_2O_3, octahedral.
Passed into $AgNO_3$ yields Ag, black, and As_2O_3.
}

Confirm by the following tests: Bloxam's (electrolysis), Bettendorff's (HCl + SnCl)As, brown, Fleitman's (Zn + 2NaHO), and
Davy's (NaHg) amalgam.†

12...

13......As_2O_3 yields yellow Ag_3AsO_3; As_2O_5 yields red brown Ag_3AsO_4.
14...... "   "   green $CuHAsO_3$;   "   "   bluish $CuHAsO_4$.
15...... "   "   .... "   "   white $MgNH_4AsO_4$.

16......As_2O_3 yields yellow As_2S_3; As_2O_5 yields yellow $As_2S_3 + S_2$.
17......Ammonium asenio-molybdate.

*The figures in this column refer to the corresponding figures in the first column
on page 41, which column is to be supplied here by the student.

†Dissolve As_2O_3 in a few drops of HCl; add a few drops of this solution to a test
tube about one fifth full of H_2O, and a small piece of sodium amalgam (HgNa), when
the following reaction will take place: $HgNa + H_2O = NaHO + \overline{H}$. This (nascent H)
combines with As to form H_3As. If a piece of filter paper moistened with silver
nitrate ($AgNO_3$) be placed over the test tube, the H_3As reduces the $AgNO_3$ to metallic
Ag (black). This is the most delicate test known.
In the above the Hg is omitted, as it takes no part in the reaction.

REACTIONS: **Sb — Atiomic weight, 122.**

*1......Sublimes at white heat.

2 .....Oxidizes to Sb_2O_3 or Sb_2O_5, insoluble in water and in dilute HNO_3.

3......Is not acted upon, unless free chlorine is present.

4......Dissolves as $SbCl_3$, decomposed by water to antimonious oxychloride, soluble in $H_2C_4H_4O_6$. Distinction from Bi.

5......Orange red precipitate Sb_2S_3, soluble in excess of the precipitant.

6......Oxidizes to Sb_2O_3 or Sb_2O_5, soluble in $H_2C_4H_4O_6$.

7......Is reduced, the globules becoming coated with oxide on C.

8 ....Abundant white fumes, antimonious oxide or antimonic oxide.

9......Yields oxide film, black, with silver nitrate, insoluble in ammonium hydrate; iodide incrustation, breathed upon, disappears transitorily; sulphide incrustation, orange.

10......Is reduced, forming a metallic film.

11 .....Evolves H_3Sb
{
Burns with bluish tinge, yielding antimonious oxide.
Deposits on porcelain plate a velvety mirror.
Mirror heated volatilizes slowly.
Mirror is not dissolved by chloride of lime.
Heated in a tube yields a reduced Sb.
Passed into silver nitrate, yields black silver antimonide (Ag_3Sb).
}

12......

13......Antimonious oxide yields black Ag_4O, insoluble in ammonium hydrate. Antimonic oxide yields white $AgSbO_3$, soluble in ammonium hydrate.

14......

15......

16......Antimonious oxide, orange red Sb_2S_3; antimonic oxide, yellowish red Sb_2S_5.

17......White precipitate.

*The figures in this column refer to the corresponding figures in the first column on page 41, which column is to be supplied here by the student.

IMPORTANT TO NOTICE.

H_2S precipitates Sb_2S_3 of 100 parts of dried sulphide $= 202.78$ parts of tartar emetic. H_2S precipitates As_2S_3 of 100 parts dried sulphide $= 80.48$ parts of pure arsenious acid.

Solubility — As_2O_3; in boiling water 1–400, cold water from 1–1000 to 1–500 of its weight.

$\overline{KSbOT}$; in boiling water 1–2, in cold water 1–14 of its weight.

FALLACIES OF REINSCH'S TEST.—Antimony, mercury, silver, bismuth, platinum, palladium, and gold are deposited upon copper, under the same conditions as arsenic.

Reinsch's test was published by Hugo Reinsch, in 1843, as an accidental discovery of As in HCl.

Marsh's test was proposed by Mr. Marsh, in 1836.

To separate As from Sb.— (1.) Take chloride of tin and fuming HCl in equal parts; As precipitates on boiling as brown metallic As (Bettendorff's. (2.) Recently precipitated sulphide of As is soluble in bisulphide of potassium, or in ammonium carbonate, while the sulphide of Sb is insoluble.

To separate Cd from As.—CdS, yellow, is not soluble in NH_4HO, but is soluble in strong HCl. As_2S_3, yellow, is soluble in NH_4HO, but is insoluble in strong HCl.

To separate As, Sn, and Sb, as sulphides.—Digest with $(NH_4)_2CO_3$, and filter $(NH_4)_3AsS_3 + (NH_4)_3AsO_3$ in solution; boil the residue with HCl, dissolves Sb_2Cl_6, $SnCl_4$.* To this add Pt + Zn, $\underline{Sb-Sn}$; wash with water. Boil with HCl, dissolves $SnCl_2$; test by $\underline{HgCl_2}$, yields white $\underline{HgCl}$. Test Sb by H_2S, yellow; test As by heating residue with $\overline{KClO_3} + HCl$, and heated in tube gives octahedral crystals.

---

*If $SnCl_4$, Sb_2Cl_6 solution is treated with Zn and HCl in the presence of Pt foil, in a Marsh's apparatus, $\overline{SbH_3}$ as a gas, and can be tested by $AgNO_3$, gives $\overline{Ag_3Sb}$. The deposit contains Sn; dissolve in HCl, and test by $HgCl_2$, as before described.

SEPARATION OF THE IRON GROUP—FIRST METHOD.

Phosphates and oxalates absent. Boil filtrate of Lead and Arsenic Groups to expel H_2S. Add a few drops of HNO_3 or $KClO_3 + HCl$, and boil an instant to oxidize Fe; immediately add NH_4Cl and NH_4HO in excess.

PRECIPITATE.

$Fe_2(OH)_6$, $Cr_2(OH)_6$, $Al_2(OH)_6$. Add NaHO or KHO, and boil for some minutes; if not boiled, Cr and Al go in solution, and may be separated by boiling afterwards, Cr.

PRECIPITATE.	SOLUTION.
Ferric hydrate, chronic hydrate. Divide into two portions.	$K_2Al_2O_4$. Add HCl to make it acid, then $(NH_4)_2CO_3$, $Al_2(OH)_6$. Heated on charcoal moistened with $Co(NO_3)_2$ blue mass.

PORTION 1.	PORTION 11.
Fe. Add HCl; test KCyS, red; K_4FeCy_6, blue.	Cr. Fuse on Pt with potassium nitrate thoroughly; dissolve in water, add a few drops of $HC_2H_3O_2$ and $Pb(C_2H_3O_2)_2$, $PbCrO_4$.

FILTRATE—METHOD 1.

Co, Ni, Mn, Zn, in solution. Add ammonium sulphide, heat gently; filter and wash the ppt. with ammonium sulphide, and also with water. Treat with cold dilute (1 to 12) HCl.

RESIDUE.	SOLUTION—METHOD 1.*
NiS, CoS. Test for Co with borax bead. Dissolve in $HNO_3 + 3HCl$.	$ZnCl_2$, $MnCl_2$. Add NaHO or KHO, and digest without warming.

METHOD 1.

Evaporate to near dryness; dilute, add HA and KNO_2.

PRECIPITATE.	FILTRATE.
$CoK_3(NO_2)_5$. H_2O.	$Ni(NO_2)_2$. Add NaHO NiH_2O_2.

METHOD 11.

Neutralize with Na_2CO_3; add KCy till the liquid is clear. Now add Br or Na ClO, boil.

PRECIPITATE.	FILTRATE.
Ni_2O_3, black.	$CoCy_2$.

PRECIPITATE.	SOLUTION.
$Mn(OH)_3$. Heated on Pt with $Na_2CO_3 + KNO_3$, green mass $NaMnO_3$; add HA, red $NaMnO_4$, or add Br.	K_2ZnO_2. Pass H_2S through ZnS, white; heated on charcoal, moistened $Co(NO_3)_2$ green mass.

NOTE.—A very little HNO_3 will oxidize the Fe; an excess makes manganic compound, that does not dissolve. NH_4Cl dissolves the manganous hydrate. NH_4HO dissolves the Ni, Co and Zn hydrates.

In all cases handle quickly and wash thoroughly, and see that the other groups are well washed out. Always test the filtrate to see that the precipitate is all down.

*SOLUTION—METHOD 11.

In HA solution. Pass H_2S through ZnS, white; must contain excess H_2S, filter; render alkaline by NH_4HO and $(NH_4)_2S$, added, MnS, flesh colored.

SEPARATION OF THE IRON GROUP—FIRST METHOD CONTINUED.

SECOND METHOD FOR FILTRATE.

Solution contains Zn, Mn, Ni, Co. Add $(NH_4)_2S$ in excess: rapidly filter (to prevent the oxidation of the sulphides), and wash the pracipitate with dilute $(NH_4)_2S$.

THE PRECIPITATE,
ZnS, MnS, NiS, CoS.

Boil the precipitate in H, $C_2H_3O_2$, and filter.

PRECIPITATE.		FILTRATE.
ZnS, NiS, CoS. Dissolve in the *least possible* quantity of HNO_3, and add NaHO in excess, and filter.		$Mn(C_2H_3O_2)_2$ Add $NH_4Cl +$ NH_4HO and $(NH_4)_2S$, flesh colored MnS; confirm as before.

PRECIPITATE.	FILTRATE.	
Ni(OH)$_2$, Co(OH)$_2$; separate as before, or dissolve in HCl; add KClO$_3$, heat until chlorous odor disappears; neutralize with Na$_2$CO$_3$, add KCy until the ppt. formed is redissolved; boil, and when cold, add a strong solution of NaClO; after standing some time, filter.	Zn(OH)$_2$. Pass H$_2$S, ZnS, white.	

PRECIPITATE.	FILTRATE.
Ni$_2$(OH)$_6$, black.	K$_6$Co$_2$Cy$_{12}$; test with bead.

Tin — Sn^{II-IV}, 118 :   TESTS.

(*1*) HNO$_3$ dissolves Sn and Sb, and forms oxides SnO$_2$ and SbO$_2$.

(*2*) NaHO, NH$_4$HO peecipitates white Sn(OH)$_2$.

(*3*) H$_2$S and $(NH_4)_2S$ precipitate dark brown SnS.

(*4*) Zn precipitates metallic Sn from its solutions.

NOTE.—Stannous chloride is distinguished from stannic chloride by its reducing action on mercuric chloride, also by its action on bichromate of potash, and by its giving with an excess of auric chloride a precipitate known as "the purple of Cassius" $(AuSnO_2(?))$.

3. The metals whose alkaline sulphides are insoluble; as, Fe, Cr, Al, Co, Ni, Mn, Zn; the Cr and Al as hydrates, the rest as sulphides; and the rare elements, uranium, iridium (thallium),

as sulphides. Beryllium, zirconium, cerium, lanthanum, didymium, titanium, tantalum, as hydrated oxides.

If there is a reason to suspect phosphates or oxalates, add to the filtrate, from the lead and arsenic group, after it is boiled, to expel the H_2S, $NH_4Cl + NH_4HO + (NH_4)_2S$ and boil the precipitate.*

CoS, NiS, MnS, ZnS, FeS, $Al_2(OH)_6$, $Cr_2(OH)_6$.

Treat with dilute HCl (1–12).

RESIDUE.	SOLUTION.†
CoS, NiS.	$MnCl_2$, $ZnCl_2$, $FeCl_2$, Al_2Cl_6, Cr_2Cl_6.

If it contains phosphoric acid, treat the $(NH_4)_2S$ precipitate as follows :

1. Dry it, but not to perfect dryness.
2. Add to the weight of P_2O_5 four times the weight of metallic tin, and mix them.
3. Heat under the hood with strong HNO_3.
4. Expel most of the HNO_3, and add H_2O.
5. The solution contains the bases as nitrates.
6. The residue contains the H_3PO_4 together with SnO_2.
7. To get the H_3PO_4, pass H_2S through the mixture. The H_3PO_4 goes into solution as the SnS is formed ; always do this to see that the work is properly done.

Now dissolve CoS, NiS in the smallest quantity of aqua regia, and add it to the solution of the other bases, and commence again as though nothing had been done with the separation, according to directions before given. The Co and Ni can be separated without adding them to the solution, as before described.

___

*Boil a part of the $(NH_4)_2S$ precipitate with Na_2CO_3, filter; add $\overline{H A}$ and $CaCl_2$ CaC_2O_4, $H_2C_2O_4$ is present. Destroy by heating in a crucible; now dissolve in HCl, and precipitate as before. The filtrate contains Sr, Ca, Ba, formerly as oxalates.

†Boil this solution, and test for phosphoric acid with ammonium molybdate $(NH_4)_2Mo_4$.

SEPARATION OF THE IRON GROUP—SECOND METHOD.

IN THE PRESENCE OF PHOSPHATES.

Determine the presence of phosphates in a separate portion. Convert the ferrous into ferric salts as before described.* Add NH_4HO and NH_4Cl, heat gently, filter, and wash thoroughly.

THE PRECIPITATE.

$Al_2(OH)_6$, $Al_2O_3P_2O_5$, $Cr_2(OH)_6$, $Fe_2(OH)_6$, $Fe_2O_3P_2O_5$, and the phosphates of Ba, Sr, Ca, and Mg. Dissolve in a little HCl, and add KHO in excess; filter.

- **PRECIPITATE.** $Fe_2(OH)_6$, and the phosphates of the alkaline earths. Digest with acetic acid, boil and filter while hot.
  - **PRECIPITATE.** Contains $Fe_2(OH)_6$, and the phosphates of the alkaline earths. Add citric acid and excess of NH_4HO; filter.
    - **RESIDUE.** $Fe_2O_3P_2O_5$. Confirm by tcsts.
    - **PRECIPITATE.** The phosphates of the alkaline earths. Dissolve in HCl; add $KC_2H_3O_2$ and Fe_2Cl_6 until the liquid remains red. Boil for some time; filter and test the filtrate for alkaline earths.
      - **FILTRATE.** Contains Fe. Test as before (not as a phosphate).

- **FILTRATE.** Contains $Al_2(OH)_6$, $Al_2O_3P_2O_5$, $Cr_2(OH)_6$. Boil for some time, and filter.
  - **PRECIPITATE.** $Cr_2(OH)_6$ green. Fuse with Na_2CO_3 + KNO_3, on Pt foil. A yellow soluble mass indicates Cr. Test as before.
  - **FILTRATE.** Contains $Al_2(OH)_6$, $Al_2O_3P_2O_5$. Add a slight excess of acetic acid, boil and filter while hot.
    - **PRECIPITATE.** $Al_2O_3P_2O_5$. Confirm as before.
    - **FILTRATE.** $Al_2(OH)_6$. Neutralize with ammonium hydrate. A white precipitate indicates Al; confirm.

* Add to a small portion of the solution strong HNO_3 and ammonium molybdate. A yellow precipitate of ammonic-phospho-molybdate indicates phosphoric acid. The precipitate comes down best when gently heated.

Iron — Fe^{IV-VI}, 56 :   TESTS.

(1) NH_4HO or $NaHO$, KHO, Na_2CO_3 with ferric salts, brownish red precipitate $Fe'''(OH)_3$.

(2) $(NH_4)_2S$, black precipitate FeS.

(3) $KCNS$ blood red solution, $Fe(CNS)_3$.

(4) K_4FeCy_6, dark blue precipitate $Fe_4(FeCy_6)_3$.

(5) Borax bead, brownish red.

(6) Easily reduced on charcoal, magnetic.

(7) Tannic acid precipitates ink or ferric tannate.

NOTE. — Distinctions between ferric and ferrous compounds.

	FERRIC COMPOUNDS.	FERROUS COMPOUNDS.
1.	Ferricyanide (K_3FeCy_6), no precipitate.	Precipitates deep blue $Fe_3(FeCy_6)_2$.
2.	Sulphocyanates, red solution, $Fe_2(CyS)_6$.	No change.
3.	Ferrocyanides precipitate blue $Fe_4(FeCy_6)_3$.	$K_2Fe(FeCy_6)$, Everitt's white.

The best solvent is HCl. In testing for iron with ferrocyanide of potassium (K_4FeCy_6), strong HCl will decompose it and give the blue; this must be borne in mind when testing.

Chromium — Cr^{IV-VI}, 52.4 :
TESTS.

(1) Gives a green bead.

(2) $Pb(C_2H_3O_2)_2$, yellow precipitate $PbCrO_4$.

(3) $HgNO_3$, red precipitate Hg_2CrO_4.

(4) $AgNO_3$, brown red precipitate Ag_2CrO_4.

(5) $NaHO$, green precipitate $Cr(OH)_3$, soluble in excess.

(6) $(NH_4)_2S$ precipitates $Cr(OH)_3$.

7

(7) The most delicate test for Cr as CrO_3, is by means of H_2O_2 (hydric peroxide) and $(C_2H_5)_2O$ (ether), giving a fine blue color, giving the reaction with one part in 40,000 of water.

Aluminium — Al''', 27.5 :

TESTS.

(1) With $Co(NO_3)_2$ on charcoal, blue mass.

(2) NH_4HO and $NaHO$ precipitates white $Al(OH)_3$, soluble in excess of $NaHO$.

(3) Na_2CO_3 and $(NH_4)_2S$ precipitates white $Al(OH)_3$; the last gives off H_2S.

NOTE. — Fe, Al and Cr form sesquioxides, R_2O_3, and may be obtained by igniting their hydrates.

Cobalt — Co'', 59 :   TESTS.

(1) Borax bead, blue, characteristic.

(2) $(NH_4)_2S$ precipitates black CoS, insoluble in HCl.

(3) Na_2HPO_4 precipitates reddish $CoHPO_4$, soluble in acids and in NH_4HO.

(4) KCyS, blue color, $Co(CyS)_2$.

(5) KNO_2 precipitates yellow $CoK_2O_2(NO_2)_4$.

(6) KCy precipitates brownish white $CoCy_2$, soluble in excess.

NOTE. — Nickel and cobalt occur together, and are separated, (1) by potassium nitrite Co, while Ni is not; (2) by cyanogen, but the process is too dangerous, except in a specially constructed laboratory.   It is, by all means, the best method of separation.*

---

* As follows:

METHOD I. — The oxides of the metals are treated with hydrocyanic acid, and then with potash, and the liquid warmed until the whole is dissolved. The reddish yellow solution is boiled to expel the hydrocyanic acid, whereupon the cobaltocyanide of potassium (K_4CoCy_6) formed in the cold, is converted into cobalticyanide, $K_6Co_2Cy_{12}$, while the nickel remains in the form of cyanide of nickel and potassium (K_2NiCy_4). Pure and finely divided red oxide of mercury is then added to the solution while yet warm, when the whole of the nickel is precipitated as oxide and cyanide, the mercury

Nickel — Ni″, 59: TESTS.

(1) Borax bead, brownish red.

(2) NaHO precipitates green $Ni(OH)_2$.

(3) $(NH_4)_2S$ precipitates black NiS, insoluble in HCl.

(4) NH_4HO precipitates green $Ni(OH)_2$, soluble in excess to a blue liquid.

(5) Na_2CO_3 precipitates green basic salt, $2NiCO_3$, $3Ni(OH)_2$.

(6) K_3FeCy_6, greenish yellow, $Ni(FeCy_6)$.

Manganese — Mn″, 55: TESTS.

(1) Borax bead, amethyst.*

(2) $(NH_4)_2S$ precipitates flesh colored MnS.

(3) NH_4HO or NaHO precipitates white $Mn(OH)_2$; turns brown in the air to Mn_2O_3.

(4) MnO_2 and $Na_2CO_3 + KNO_3$, fused on platinum foil, greenish blue mass (sodic manganate), Na_2MnO_4.

NOTE. — Manganese is most easily and certainly identified through oxidation by several methods, each method giving a colored product.†

---

taking its place in the solution. The filtered solution contains all the cobalt as cobalticyanide of potassium. This is known as Liebig's method.

METHOD II (ROSE). — This depends upon the fact, that CoO in acid solution is converted by Br into Co_2O_3; whereas, with Ni, this change does not take place. The sulphides are dissolved in "aqua regia" and the solution diluted, and then Br added to saturation; $BaCO_3$ is now added in excess. Let stand for ten or twelve hours, and well shaken up from time to time. The precipitate Co_2O_3 and $BaCO_3$ is filtered, and washed with cold water; the filtrate contains the Ni without a trace of Co. The precipitate is boiled with HCl, which converts Co_2O_3 into CoO, and the $BaCO_3$ into $BaCl_2$. The $BaCl_2$ is precipitated by H_2SO_4, and the Co from the filtrate by KHO. The use of the $BaCO_3$ is to precipitate Co_2O_3.

Add to a solution of Co and Ni, $NH_4Cl + NH_4HO$ with K_3FeCy_8, which gives a blood red color and indicates Co. If Ni be present and the solution is boiled, a copper red precipitate forms; if only Co is present, a dirty green on boiling. This is a very ready means of detecting the presence of Co and Ni.

*A delicate test is to oxidize the bead with a small crystal of KNO_3, and get a permanganate compound.

†This test can be nicely shown by the electrolysis of a manganese compound in presence of HNO_3, when but a trace gives a pink solution, a very delicate test.

Zinc — Zn″, 65 :   TESTS.

(*1*) Changes color on charcoal, yellow while hot; on charcoal with Co(NO_3)$_2$, green mass.

(*2*) NH$_4$HO or NaHO precipitates white Zn(OH)$_2$, soluble in
excess.

(*3*) H$_2$S. precipitates white ZnS in presence NaC$_2$H$_3$O$_2$.

(*4*) (NH$_4$)$_2$S precipitates white ZnS, soluble in HC$_2$H$_3$O$_2$
(acetic acid).

(*5*) K$_2$CO$_3$ precipitates in white basic carbonate, Zn$_5$(CO$_3$)$_2$,
(HO)$_6$, soluble in NH$_4$HO.

(*6*) KCy precipitates white ZnCy$_2$, soluble in excess.

(*7*) K$_4$FeCy$_6$ precipitates white Zn$_2$FeCy$_6$.

(*8*) K$_3$FeCy$_6$ precipitates yellowish Zn$_3$(FeCy$_6$)$_2$.

NOTE. — ZnS is the only sulphide that is precipitated white. In
general terms, we can say one alkali will replace another in the
tests, and for separation. This is regarded by many as the most
important group in the whole course of qualitative separation; it
therefore demands a proportionate amount of care and study.

4. This group consists of the metals, Ba, Sr, Ca, Mg, whose
carbonates are insoluble in water; if NH$_4$Cl be present, Mg is
left for the next group, also, the rare elements, lithium, cæsium
and rubidium.

SEPARATION OF THE ALKALINE EARTHS.

Boil the ammonium sulphide precipitate from the last group to decompose
ammonium sulphide.

Add $NH_4Cl + NH_4HO + (NH_4)_2CO_3$.

PRECIPITATE — METHOD I.*	FILTRATE.
$BaCO_3$, $SrCO_3$, $CaCO_3$. Dissolve in HNO_3, evaporate to dryness, pulverize finely, and treat with *absolute* alcohol (C_2H_5OH).	$MgCO_3$.†

RESIDUE.	SOLUTION.
$Ba(NO_3)_2$, $Sr(NO_3)_2$. Dissolve in H_2O and add $(NH_4)_2CO_3$.	$Ca(NO_3)_2$, Add $(NH_4)_2$ C_2O_4, CaC_2O_4, white. See properties of this ppt. before d'scrib'd. Flame, red.

PRECIPITATE.	
$BaCO_3$, $SrCO_3$. Dissolve in HCl, evaporate to dryness, and again treat with absolute alcohol.	

RESIDUE.	SOLUTION.
$BaCl_2$. Test with flame.	$SrCl_2$. Test with flame, carmine.

* and † See Method II on the next page.

SEPARATION OF THE ALKALINE EARTHS.

*PRECIPITATE — METHOD II.	FILTRATE.
$BaCO_3$, $SrCO_3$, $CaCO_3$. Add H, $C_2H_3O_2$; solution as acetates; now add K_2CrO_4.	$MgCO_3$.†

PRECIPITATE.	FILTRATE.
$BaCrO_4$. Dissolve in HCl; flame green; add H_2SO_4, white $\underline{BaSO_4}$, insoluble in acids.	$SrCrO_4$, $CaCrO_4$. Add $(NH_4)_2CO_3$.

PRECIPITATE.

$SrCO_3$, $CaCO_3$; wash this well to remove K_2CrO_4, and dissolve in acetic acid.

SOLUTION.

$Sr\overline{A}$ and $Ca\overline{A}$; now add K_2SO_4 (1 to 200 H_2O).

PRECIPITATE.	FILTRATE.
$SrSO_4$.	$CaSO_4$, add $(NH_4)_2C_2O_4$, $\underline{CaC_2O_4}$, white.

†We now have in the filtrate, or may have, $MgCO_3$, K_2CO_3, Na_2CO_3, Li_2CO_3.

Take a small part, and test for Mg by Na_2HPO_4; if present, a white $\underline{MgNH_4PO_4}$. Evaporate to dryness; heat to expel the NH_4HO salts; now add $H_2C_2O_4$ to convert into oxalates, and again heat to convert the Mg to oxide, and the other bases to carbonates. Dissolve in water; before you dissolve in water, test Li by flame reaction.

SEPARATION OF THE ALKALIES.

RESIDUE.	SOLUTION.	
MgO. ·· Test with Co(NO$_3$)$_2$ on charcoal, pink.	K$_2$CO$_3$, Na$_2$CO$_3$. Treat with HCl; add C$_2$H$_5$OH + PtCl$_4$.	
	PRECIPITATE.	FILTRATE.
	2KCl, PtCl$_4$, yellow crystalline.	2NaCl, PtCl$_4$. Test by flame, yellow.

Test the original solution for NH$_4$HO by boiling with NaHO, and testing with litmus or by HCl. Nessler's test (HgK$_2$I$_4$) gives brownish Hg$_2$N$_2$I$_2$, 2H$_2$O; this test is for traces. For other delicate tests, see Second Supplement to Watt's Dictionary of Chemistry, page 59; and for the most delicate test, see Third Supplement, Part I, page 73 of the same work.

NOTE.—Test for K in the original solution; for it may be necessary to add a salt of it—KClO$_3$ or KNO$_3$, as an oxidizing agent in some of the groups. Na is everywhere, and it is not generally added to a mixture of the solutions for qualitative separation.

GROUP OF ALKALINE EARTHS—BA, SR, CA, MG.

SULPHATES.		CARBONATES.	
BaSO$_4$	Heavy spar.	BaCO$_3$......	Witherite.
SrSO$_4$	Celestine.	SrCO$_3$......	Strontianite.
CaSO$_4$......	Gypsum, selenite.	CaCO$_3$......	Chalk, limestone.
	CaCO$_3$, MgCO$_3$, dolomite.		

NUMBER OF PARTS OF WATER REQUIRED TO DISSOLVE ONE PART OF THE SALT.

HYDRATES.		SULPHATES.	
Ba	15.	Ba	(?) 200,000.
Sr	60.	Sr.	7,000.
Ca	700.	Ca	400.
Mg	6,000.	Mg	3.

NAME.	SYMBOL.	ATOMIC WEIGHT.	SPECIFIC GRAVITY.
Barium	Ba''.	137.	4.
Strontium	Sr''.	87.5	2.5
Calcium	Ca''.	40.	1.58
Magnesium	Mg''.	24.	1.74

NOTICE.

1. $(NH_4)_2CO_3$ precipitates Ba, Sr, Ca *completely* from their solutions as carbonates, but it precipitates Mg *only partially*, and this may be prevented, at least for a time, if NH_4Cl is present in the solution.

2. The above is the order of their electro positive series.

3. Their atomic weights decrease in this order.

4. Their specific gravities decrease in this order, except Mg.

5. Notice the solubility of their hydrates.

6. Notice the solubility of their sulphates.

7. The metals decompose water (Mg only slowly).

Barium — Ba'', 137 :   TESTS.

(1) Chlorides give *greenish yellow* flame.

(2) $(NH_4)_2CO_3$ precipitates white $BaCO_3$.

(3) H_2SO_4 precipitates white $BaSO_4$, insoluble in acids.

(4) NaOH precipitates white $Ba(OH)_2$, soluble in boiling water.

(5) $K_2Cr_2O_7$ precipitates yellow $BaCrO_4$.

(6) Na_2HPO_4 precipitates white $BaHPO_4$.

(7) Hydrofluosilicic acid ($SiHFl_4$, $2HFl$) precipitates white $SiFl_4$, $BaFl_2$.

(8) $(NH_4)_2C_2O_4$ precipitates white BaC_2O_4.

Strontium — Sr″, 87.5 : TESTS.

(1) Flame test, crimson.

(2) $NaOH$, NH_4HO, Na_2CO_3, $(NH_4)_2CO_3$, Na_2HPO_4, H_2SO_4 and $(NH_4)_2C_2O_4$ give the same tests as with Ba.

(3) But (5) and (7) above do not give any precipitates for Sr, while they do with Ba.

(4) The sulphates and carbonates are more soluble than Ba compounds.

Calcium — Ca″, 40 : TESTS.

(1) Flame, yellowish red.

(2) $NaOH$, NH_4HO, Na_2CO_3, Na_2HPO_4, $(NH_4)_2C_2O_4$, H_2SO_4, same as Ba.

NOTE.—The flame reaction will distinguish this from the other members of the group. A solution of $CaSO_4$ (not too little) precipitates Ba *immediately*, Sr *after some time*, Ca *not at all*. They can also be identified by their flame reactions as before described.

Magnesium — Mg″, 24 : TESTS.

(1) With $Co(NO_3)_2$, flesh colored mass on charcoal.

(2) NH_4HO and $NaHO$ give white precipitate $Mg(OH)_2$, soluble in NH_4Cl.

(3) Na_2CO_3, white precipitate $MgCO_3$, soluble in NH_4Cl.

(4) Add NH_4Cl and NH_4HO to the solution, and now add Na_2HPO_4, white precipitate $MgNH_4PO_4$.

It is well to notice in this test Na_2HPO_4 *alone* will not give a precipitate in dilute solutions; now NH_4HO precipitates but one-

8

half of the Mg. $2MgSO_4 + 2NH_4HO = Mg(OH)_2 + (NH_4)_2SO_4$, $MgSO_4$. The addition of NH_4Cl prevents formation of $Mg(OH)_2$. The NH_4HO makes the precipitate more insoluble than in water. $MgNH_4PO_4$ is soluble in 15,000 parts of pure water, or in 40,000 parts of water containing NH_4HO.

5. This group consists of the alkali metals; their chlorides, sulphides and carbonates are *soluble in water*, and they are *not* precipitated; K, Na, Li, NH₄ and the rare metals, cæsium, rubidium and lithium belong in this class.

Potassium — K', 39 : TESTS.

(*1*) By the flame, violet.

(*2*) $PtCl_4$ precipitates yellow $(KCl)_2PtCl_4$.

(*3*) $NaHC_4H_4O_6$ precipitates white $KHC_4H_4O_6$.

(*4*) Nitrophenic acid $(HC_6H_2(NO_2)_3O)$ precipitates yellow $KC_6H_2(NO_2)_3O$; *when dried and heated, explosive.*

Sodium — Na', 23 : TESTS.

(*1*) Flame, yellow.

(*2*) $PtCl_4$ forms red $(NaCl)_2PtCl_4$, very soluble.

(*3*) Potassium metantimoniate $(KSbO_3)$ precipitates white $NaSbO_3$; the reagent must be made when required, as it is not permanent in solution.*

Lithium — Li', 7 : TESTS.

(*1*) Flame, carmine red.

(*2*) Na_2HPO_4 precipitates white Li_3PO_4.

(*3*) Nitrophenic acid precipitates yellow $LiC_6H_2(NO_2)_3O$.

---

*Made by fusing antimonic acid with a large excess of potassium hydrate; then dissolving, filtering, evaporating and digesting, hot, in a concentrated solution, with a large excess of potassium hydrate, in a silver dish, decanting the alkaline liquor, and stirring the residue to granulate; now, dry. This must be kept dry. Dissolve a small portion, when required, as the solution changes to the dibasic metantimoniate, which does not precipitate sodium. The reagent can not be used in acid solutions.

Ammonium has been sufficiently described.

Cæsium — Cs', 133:   TEST.

(1) Flame, violet red.

Rubidium — Rb', 85:   TEST.

(1) Flame, violet red.

Cæsium and rubidium are best identified by the spectroscope.

NAME.	SYMBOL.	ATOMIC WEIGHT.	SPECIFIC GRAVITY.
Lithium	Li.	7.	.589
Potassium	K.	39.	.865
Sodium	Na.	23.	.985

If we take them in the electro positive series — Cs, Rb, K, Na, Li, their atomic weights decrease in this order (see above); their specific gravities decrease (except K), their fusing points rise, and the solubility of their carbonates lessen in this order.

It will be well for the student to remember that

1. H_2SO_4 makes *sulphates* of the metals; HCl and "aqua regia" $(3HCl + HNO_3)$ makes *chlorides;* HNO_3 makes *nitrates*, except Sb, Sn, oxides, of the metals.

2. *Each group reagent will precipitate the metals of preceding groups;* as, for instance, the chlorides of the first group are also insoluble as sulphides with the second and third groups, and as carbonates with the fourth group. The second and third group metals form insoluble carbonates as well as those of the fourth group. The first and second groups can be worked together, but after that *each group must be completely removed before testing for the next group,* and the *filtrate always tested with the group reagent.* The groups are arranged in the order in which you would work out an unknown body.

3. If a salt is given for analysis, dissolve in water; Bi and Sb may form oxychlorides.

4. If a metal, dissolve in acid; Pt and Au require "aqua regia."

An alloy is a mixture of metals, which may be either *natural* or *artificial;* when one of the metals of an alloy is Hg, it is called an *amalgam.*

ARTIFICIAL ALLOYS.	COMPOSITION.
Gun metal	Cu 90, Sn 10 parts.
Bronze	Cu 91, Sn 6, Pb 1 part.
Brass	Cu 28, Zn 34 parts.
Bell metal	Cu 72, Sn 22 parts.
Speculum metal	Cu 75, Sn 25 parts.
German silver	Cu 100, Zn 60, Ni 40 parts.
Pewter	Zn 12, Sb 1, Cu (a little).
Plumber's solder	Pb 2, Sn 1 part.
Fine solder	Pb 1, Sn 2 parts.
Type metal	Pb 73, Sb 17, Sn 10 parts.
Standard silver	Ag 90, Cu 10 parts.
Britannia metal	Sn 9, Sb 1, also Cu, Zn, Bi.

If we take the first (gun metal) and dissolve it in HNO_3, we convert the Sn into SnO_2, and we have $Cu(NO_3)_2$ in solution; the Cu may be precipitated while hot by NaHO. The student will analyze some of these compounds, and test his knowledge of chemistry by selecting the proper solvent, and method of treatment.

Zettnow has arranged a scheme without the use of H_2S or $(NH_4)_2S$.

ZETTNOW'S METHOD OF SEPARATION.

Add hydrochloric acid to the solution, wash, and filter.

PRECIPITATE. Boil with water and filter.			FILTRATE. Add excess of dilute H_2SO_4, and wash on filter.

| SOLUTION. Add H_2SO_4. | RESIDUE. Treat with $(NH_4)HO$. | | PRECIPITATE. Agitate with considerable cold water and filter. | | | To $\frac{1}{3}$ add BaH_2O_2 and boil. | | | | FILTRATE. Divide the solution into two unequal parts, $\frac{1}{4}$ and $\frac{3}{4}$. |

| PRECIPITATE. Pb. | SOLUTION. Add HNO_3. PRECIPITATE. Ag. | RESIDUE turns gray or black. Hg. | FILTRATE. $(NH_4)_4C_2O_4$, digest and filter. | RESIDUE. $(NH_4)_4C_2H_3O_6$, and $(NH_4)_4C_2H_3O_6$. |

In this scheme regard is had to the following substances in aqueous solution:

I.   PbO, Ag_2O, HgO.
II. CaO, BaO, SrO.
III. $(NH_4)_2O$, Na_2O, K_2O.
IV. As_2O_3, As_2O_5, Sb_2O_3, Sb_2O_5, SnO, SnO_2, Hg_2O, CuO, CdO, Bi_2O_3.
V.   FeO, Fe_2O_3, Cr_2O_3, Al_2O_3.
VI. MnO, MgO, CoO, NiO.
VII. ZnO.

N. B.—To test for zinc mix a portion of the original solution with HCl, H_2SO_4, filter, add $Na\,HO$ in excess, and boil. Make $(NH_4)_2CO_3$ and NH_4Cl to filtrate. boil until all odor of $(NH_4)HO$ is expelled, and filter. Add K_4FeCy_6 to solution; a cloud or precipitate indicates Zn.

CHAPTER IV.

SEPARATION OF THE ACIDS.

The examination for the bases should always precede the examination for the acids, except H_2SO_4, HCl and HNO_3.

NOTICE.

1. We may find while searching for the bases, As_2O_3, Sb_2O_3, CO_2, SiO_2, CrO_3, H_2S, H_3PO_4, $H_2C_2O_4$, H_2SO_3, and some others.

2. S, Se and Te give a hepar on silver coin, also PH_2 is self-enkindling.

3. Deflagration on charcoal indicates *chlorates, nitrates, permanganates, chromates, bromates* and *iodates*.

4. Certain bases form with certain acids, compounds insoluble in water — H_2SO_4 forms with Ba, Sr, Ca and Pb, corresponding sulphates; HCl forms with Ag and Hg, corresponding chlorides; and P_2O_5 forms with Ba, Sr and Ca, corresponding phosphates.

5. Heating separates the acids into two great groups — organic and inorganic; all the organic acids blacken, when heated to redness, except *acetic* and *formic* acids, and their salts.

Acid solutions of their salts are tested in the *wet* and *dry* ways.

IN THE DRY WAY.

In the *dry way,* evaporate to dryness, and heat in a small tube with four times its weight of strong sulphuric acid.

NOTICE. — The chlorides of Pb, Ag, Hg, Sn and Sb, and the sulphide of As, are not decomposed by H_2SO_4, or but slowly; CO_2, SO_2, H_2S, HCl, HF and HNO_3 are liberated in the free state, while HI, HBr, HCN, CrO_3, $HClO_3$ and $H_2C_2O_4$ decompose.; the gas may be either colored or colorless.

COLORED GAS EVOLVED.	COLORLESS GAS EVOLVED.
I......Violet.	With an odor—
Br.....Reddish.	HC₂H₃O₂....of vinegar.
NO₂;...Reddish.	H₂S .....of rotten eggs.
Cl₂O₅...Greenish yellow.	SO₂ ...of burning sulphur.
	HCy .....of peach kernels.
	HF.........Pungent, etches glass.
	H₂C₄H₄O₆ ...of burnt sugar.;ᵢₙ
	Without an odor—
	CO₂, H₂C₂O₄, CO.

NOTE.—The reactions may be modified, if there is more than one acid in the substance to be examined.

IN THE WET WAY.

Dilute sulphuric acid in the *wet way*, with zinc, colors the fluid—

Violet or lilac, titanic acid (TiO_2).

Blue, tungstic and vanadic acids (WO_3, V_2O_5).

Blue, then green and brown, molybdic acid (MoO_3).

Blue, then muddy or brown, tantalic and niobic acids (TaO_2, Nb_2O_5).

Green, chromic acid (CrO_3).*

To get a solid, containing an acid, in solution, it must be finely pulverized. See if H_2O, HCl, HNO_3 or "aqua regia" (3HCl + HNO_3) will dissolve it; if not, fuse it with four times its weight of the double carbonates of potash and soda, and get it in solution in water or hydrochloric acid. The hydrochloric acid will now dissolve the metals, and render the SiO_2 insoluble, if present.

---

*Take a small beaker, add the supposed acid; now add sulphuric acid and zinc, as in making hydrogen. The nascent hydrogen reduces the acid, and changes the color of the solution as above indicated.

INORGANIC ACIDS.

The inorganic acids consist of the following groups:

1. BaCl$_2$ in the presence of HCl precipitates sulphuric (H$_2$SO$_4$) and hydrofluosilicic (H$_2$SiF$_6$) acids.

Sulphuric acid and sulphur:.

TESTS.

(*1*) Gives a hepar with Mg or Na on silver coin. To make the hepar test, take a broken test tube, draw it out to a quill shaped point, and put a small piece of Mg or Na, and some of the dry substance to be tested, together in the tube; now heat, and when the Mg or Na burns, break off the end of the glass on a silver coin, and add one or two drops of water; a black stain indicates S.

(*2*) To detect S in the albumen, take a solution of acetate of lead, add sodium hydrate until clear; add the albumen to the solution, and boil (black colored sulphide).

(*3*) With a solution of nitroprusside of sodium, added to an *alkaline* solution of sulphur, is of a violet color, very delicate.

(*4*) Alkaline earths precipitate white sulphates (best in this order), Ba, Sr, Ca and Pb (quantitative).

Hydrofluosilicic acid: TESTS.

(*1*) BaCl$_2$ precipitates BaSiF$_6$.

(*2*) KCl precipitates gelatinous K$_2$SiF$_6$.

(*3*) H$_2$SO$_4$ decomposes it; liberated $\overline{\text{HF}}$ etches glass.

2. BaCl$_2$ in *neutral solutions* precipitates phosphoric acid (H$_3$PO$_4$), boric acid (H$_3$BO$_3$), oxalic acid (H$_2$C$_2$O$_4$), hydrofluoric acid (HF), carbonic acid (H$_2$CO$_3$), silicic acid (H$_4$SiO$_4$), sulphurous acid (H$_2$SO$_3$), hyposulphurous acid (H$_2$S$_2$O$_3$), arsenious acid (H$_3$AsO$_3$), arsenic acid (H$_3$AsO$_4$), iodic acid (HIO$_3$), and chromic acid (H$_2$CrO$_4$).

Phosphoric acid: TESTS.

(*1*) $BaCl_2$ precipitates white $BaHPO_4$.

(*2*) "Magnesia mixture" precipitates white $MgNH_4PO_4$.

(*3*) $Pb(C_2H_3O_2)_2$ precipitates white $Pb_3(PO_4)_2$.

(*4*) $AgNO_3$ precipitates yellow Ag_3PO_4; these four are soluble in HNO_3.

(*5*) $(NH_4)_2MoO_4 + HNO_3$ gives a yellow precipitate of *ammonium phosphomolybdate* of a *variable composition*.

Pyrophosphoric acid with $AgNO_3$ precipitates white $Ag_4P_2O_7$.

Metaphosphoric acid with $AgNO_3$ gives a white gelatinous precipitate of $AgPO_3$.

Boric acid: TESTS.

(*1*) $BaCl_2$ precipitates white $Ba(BO_2)_2$.

(*2*) $AgNO_3$ gives a yellowish precipitate; in dilute solutions, Ag_2O is precipitated; in acid salts, precipitates white $AgBO_2$.

(*3*) When burnt with C_2H_5HO, gives a green flame.

Oxalic acid—The oxalic acid is here placed with the inorganic acids.

             TESTS.

(*1*) Sulphuric acid breaks it up into CO and CO_2; CO burns with a blue flame.

(*2*) $BaCl_2$ precipitates white BaC_2O_4.

(*3*) $CaCl_2$ precipitates white CaC_2O_4.

(*4*) $AgNO_3$ precipitates white $Ag_2C_2O_4$, all soluble in HNO_3.

Hydrofluoric acid: TESTS.

(*1*) Sulphuric acid acting upon fluorides evolves $\overline{HF}$, which etches glass.

(*2*) $BaCl_2$ precipitates white BaF_2, soluble in HCl.

(*3*) $CaCl_2$ precipitates transparent gelatinous CaF_2.

Carbonic acid : TESTS.

(1) Carbonates of the metals are generally insoluble in water, excepting the alkalies and some bicarbonates.

(2) All carbonates effervesce with hydrochloric acid, giving off free CO_2.

(3) CO_2 produces in a drop of $Ca(HO)_2$ a white turbidity of $CaCO_3$.

(4) $BaCl_2$ precipitates white $BaCO_3$.

Silicic acid : TESTS.

(1) $BaCl_2$ precipitates white Ba_2SiO_4.

(2) HCl precipitates white gelatinous H_4SiO_4; if evaporated to dryness, becomes insoluble SiO_2.

(3) Fuse in a Pt crucible with four times its weight of $KNaCO_3$; the SiO_2 becomes *soluble*.

(4) Make a bead of microcosmic salt; the SiO_2 floats on the bead undissolved, known as a *skeleton* bead.

The remaining six acids of this group are *precipitated* or *decomposed* by the group reagents for the *bases*.

H_2SO_3, decomposed by HCl, gives $\overline{SO_2}$. $H_2S_2O_3$, decomposed by HCl, gives $\overline{SO_2}$ and S.

H_3AsO_3 is precipitated by H_2S as yellow As_2S_3. H_3AsO_4 is precipitated by H_2S as yellow As_2S_3. HIO_3 is decomposed by H_2S and forms an iodide, while S separates.

H_2CrO_4 is precipitated by $(NH_4)_2S$ as $Cr_2(OH)_6$.

The *Cr* acids and the *As* acids have been described in the *tests for the bases*.

Sulphurous acid : TESTS.

(1) $BaCl_2$ precipitates white $BaSO_3$, soluble in HCl.

(2) Add Zn + HCl of $\overline{H_2S}$, odor.

9

(*3*) HCl decomposes it as $\overline{SO_2}$, which may be recognized (*a*) by the odor, (*b*) by bleaching, (*c*) by being a powerful reducing agent. It does this by taking the oxygen away from the body, and itself being changed to *sulphuric acid.*

Sulphuric acid.— H_2SO_4 + $BaCl_2$ precipitates white $BaSO_4$, *insoluble* in HCl or HNO_3.

Sulphurous acid.— H_2SO_3 + $BaCl_2$ precipitates white $BaSO_3$, *soluble* in dilute HCl.

Thiosulphuric or hyposulphurous acid.— $H_2S_2O_3$ + $BaCl_2$ precipitates white BaS_2O_3, *soluble* in dilute HCl.

NOTICE. — This reaction separates *sulphates* from *sulphites* and *thiosulphates.* $CaCl_2$ precipitates sulphites, but does not precipitate thiosulphates. Free H_2SO_4 can be detected in the presence of a sulphate; the fluid is mixed with a very little cane sugar, and the mixture evaporated to dryness in a porcelain dish at 100° C.; if *free* sulphuric acid is present, a black residue remains; other acids do not decompose cane sugar in this way.

Iodic acid: TESTS.

(*1*) $BaCl_2$ precipitates white $Ba(IO_3)_2$.

(*2*) $AgNO_3$ precipitates white $AgIO_3$, soluble in ammonia.

(*3*) SO_2 and H_2S decompose it; test for free I by starch or with bisulphide of carbon.

(*4*) If strongly heated, $\overline{O}$; in some cases, I.

NOTICE. — If an iodate and an iodide are mixed, and treated with almost any strong acid, as follows, $5HI + HIO_3 = 6I + 3H_2O$. Alcohol precipitates *potassium iodate* from an aqueous solution, an approximate separation from iodide.

3. Acids precipitated by $AgNO_3$ and not by $BaCl_2$ — hydrochloric acid (HCl), hydrobromic acid (HBr), hydriodic acid (HI), hydrocyanic acid (HCN), hypochlorous acid (HClO), nitrous acid (HNO_2), hydrosulphuric or sulphuretted hydrogen (H_2S).

Hydrochloric acid: TESTS.

(1) $AgNO_3$ precipitates a curdy white AgCl, soluble in NH_4HO.

(2) $MnO_2 + H_2SO_4 + RCl = \overline{Cl}$.

(3) Bead $NaNH_4HPO_4 + CuO$; now expose to reducing flame, blue to purple.

(4) Solid chloride is heated with $K_2Cr_2O_7 + H_2SO_4$; brown gas changes to blood red liquid chlorochromic acid (CrO_2Cl_2); now add NH_4HO, changes to $(NH_4)_2CrO_4$, yellow; now add an acid, reddish yellow $(NH_4)_2Cr_2O_7$.

(5) Free Cl can be detected by changing ferrous to a ferric salt; test by a sulphocyanide. See page 49, Fe tests.

(6) $3HCl + HNO_3 = $ "aqua regia," dissolves Au, Pt, etc.

$$3HCl + HNO_3 = \begin{cases} NOCl_2 + Cl \\ NOCl + Cl_2 \end{cases} + 2H_2O.$$

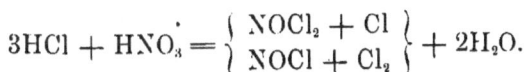

The solvent action is due to the free Cl; some authorities say $NOCl_2$ or $NOCl$, in the presence of H_2O, also liberates Cl.

Hydrobromic acid: TESTS.

(1) $AgNO_3$ precipitates pale yellow AgBr, soluble in HCl.

(2) $MnO_2 + H_2SO_4 + RBr = \overline{Br}$; recognized by odor.

(3) Cl passed through a solution of a bromide decomposes it, and liberates Br and colors it yellow.

(4) Bisulphide of carbon, or chloroform or ether gives, with free Br in solutions, reddish yellow; free Br color starch solution orange yellow.

Hydriodic acid: TESTS.

(1) $AgNO_3$ precipitates yellow AgI.

(2) $HgCl_2$ precipitates reddish yellow HgI_2.

(3) $MnO_2 + H_2SO_4 + RI = \overline{I}$.

(4) Cl water or gas liberates $\overline{\text{I}}$, but an *excess* causes ICl_3, colorless, and gives no blue with starch.

(5) Add to a solution containing RI some HCl, then KNO_2; a dark brown color indicates I.

(6) Free I colors cold starch solutions a beautiful blue, disappears on warming. Free I is soluble in CS_2 and in $CHCl_3$, with characteristic reddish color.

(7) The union of iodine and starch is probably an example of molecular adhesion rather than of union within the molecule. The bead $NaNH_4HPO_4 + CuO$ gives emerald green.

NOTICE. — HI and HIO_3 give off I, but only the *last* gives off O. It is well to bear in mind that Cl will liberate Br, and in turn Br will liberate I. For the estimation of I in the presence of Cl, see Watt's Dictionary, Second Supplement, page 674; and for a quantitative method of separating each ingredient in a mixture of all of them, see Roscoe & Schorlemmer, Vol. I, page 162, also Galloway's Qualitative Analysis, page 197.

Hydrocyanic acid :   TESTS.

(1) $AgNO_3$ precipitates white AgCN, insoluble in HNO_3; AgCN is decomposed by heat to Ag, while AgCl is *not decomposed* by heat alone.

(2) Take a few drops of HCN, and add one or two drops of yellow $(NH_4)_2S$, and heat on the water bath until colorless, and free from H_2S; add one drop of HCl, and one drop of Fe_2Cl_6, blood red color (see iron salt).

(3) Take a little KHO and $FeSO_4$, and warm the mixture for a short time; a very little Fe_2Cl_6 and the whole *slightly* acidulated with HCl to dissolve the ferrous and ferric hydrates, when Prussian blue will appear, if HCN is present — *(a)* $2KCy + FeSO_4 = FeCy_2 + K_2SO_4$; *(b)* $FeCy_2 + 4KCy = K_4FeCy_6$; *(c)* $3K_4FeCy_6 + 2Fe_2Cl_6 = \underline{Fe_4(FeCy_6)_3} + 12KCl$.

(*4*) A solution of HCN is added in excess to one part of KHO and three parts of finely pulverized HgO, which dissolves in a solution of the alkalies *only in the presence of HCN.* This is regarded by Fresenius as a *positive test for HCy.*

Note. — Neither (3) or (4) will give these tests for $HgCy_2$; the $HgCy_2$ must be treated with H_2S, when HgS and HCN remain in solution, and can be tested.

(*5*) (Schœnbein) moisten paper with guaiacum; when dry, moisten with $CuSO_4$; when this is brought in contact with HCN in air or water, it is colored blue. Notice many of the guaiacum compounds sold in the shops are not good.

Hypochlorous acid: TESTS.

(*1*) $AgNO_3$ precipitates white AgCl.

(*2*) $MnCl_2$ precipitates dark brown $\underline{Mn(OH)_2}$.

(*3*) $Pb(NO_3)_2$ white precipitate changes to brown $\underline{PbO_2}$.

Notice. — Its salts are very unstable, whether solid or in solution; the weakest acids, as well as the carbonic acid of the air, causes a liberation of Cl; indigo and litmus are decolorized by it on the addition of an acid.

Nitrous acid: TESTS.

(*1*) $AgNO_3$ precipitates white $AgNO_2$, soluble in excess.

(*2*) $FeSO_4$ gives a black coloration of $2FeSO_4$. NO.

Note. — It can be easily oxidized to HNO_3, and gives the common tests for nitrates, but differs in this respect — a nitrite decolors a $K_2Mn_2O_8$ solution, acidulated with H_2SO_4, distinct from nitrate. Nitrites when decomposed in dilute cold solutions by acetic acid or very dilute sulphuric acid, *instantly liberate iodine* from iodides; nitrates do this but slowly, and it must be in quite strong solutions.

Hydrosulphuric acid: TESTS.

(*1*) AgNO$_3$ precipitates black Ag$_2$S, insoluble in dilute acids.

NOTE. — See Pb in the bases for the best tests, also, the test for
sulphur. HCl and H$_2$SO$_4$ decompose most sulphides with $\overline{\text{H}_2\text{S}}$;
as, FeS $+$ H$_2$SO$_4$ $=$ FeSO$_4$ $+$ $\overline{\text{H}_2\text{S}}$.

It has been sufficiently described as a group reagent in the
separation of the bases.

4. Acids not precipitated by any reagent — nitric acid (HNO$_3$),
chloric acid (HClO$_3$), and perchloric acid (HClO$_4$).

Nitric acid: TESTS.

(*1*) To a solution containing HNO$_3$, add one-fourth of its
volume of strong H$_2$SO$_4$, now let it cool; a crystal
of ferrous sulphate is added, and a few drops of the
nitrate to be tested, a reddish brown ring of
2(FeSO$_4$)NO surrounds the crystal.

(*2*) If you liberate HNO$_3$ by H$_2$SO$_4$, and add a little Cu, red
fumes of N$_2$O$_4$ arise as follows : NO $+$ O $=$ NO$_2$.

(*3*) Nitrates deflagrate on charcoal.

(*4*) Brucin dissolved in sulphuric acid gives blood red.

(*5*) Phenol (C$_6$H$_5$OH), red brown color, gives nitrophenol or
picric acid, C$_6$H$_5$(NO$_2$)$_3$OH.

An aqueous solution of ferrous sulphate turns red brown on
the addition of a nitrous solution, even if *no* sulphuric acid be
added. This reaction will distinguish nitrites from nitrates (see
water analysis).

Chloric acid: TESTS.

(*1*) Sulphuric acid decomposes chlorates — Cl$_2$O$_4$, *greenish
yellow gas, explosive ; use small quantities.*

(*2*) Chlorates as KClO$_3$ give off $\overline{\text{O}}$ easily by heat as in
making oxygen.

(*3*) If a mixture of chlorate and KCy be heated on Pt foil, violet deflagrations ensue ; *use small quantities.*

(*4*) H_2SO_4 in the presence of a chlorate, bleaches indigo.

Perchloric acid: TESTS.

(*1*) Sulphuric acid does not act upon perchlorates in the cold ; when heated, $\overline{HClO}$ gives *white fumes ;* compare (1) above.

(*2*) The same as (2) above.

(*3*) KCl in strong solutions precipitates white $KClO_4$.

(*4*) Perchlorates + HCl do *not* decolorize indigo solutions.

ORGANIC ACIDS.

The common organic acids consist of the following groups :

1. Acids precipitated by $CaCl_2$ in the cold, or on boiling — tartaric acid ($H_2C_4H_4O_6$ or $H_2\overline{T}$), citric acid ($H_3C_6H_5O_7$ or $H_3\overline{Ci}$), and oxalic acid ($H_2C_2O_4$ or $H_2\overline{O}$. See oxalic acid on page 66.*

Tartaric acid: TESTS.

(*1*) When heated, gives an odor of burnt sugar.

(*2*) $CaCl_2$ precipitates white $CaC_4H_4O_6$.

(*3*) KCl precipitates white crystalline $KC_4H_5O_6$.

(*4*) $Ca(OH)_2$ precipitates white $CaC_4H_4O_6$.

(*5*) $AgNO_3$ precipitates white $Ag_2C_4H_4O_6$, soluble in HNO_3 and in NH_4HO.

(*6*) When heated with sulphuric acid, it blackens from the separation of carbon, and SO_2, CO and CO_2 are evolved.

---

*Tartaric and citric acids prevent the precipitation of iron and other heavy metals by the alkalies, soluble double salts being formed. Alkaline citrates are sparingly soluble in hot, and less soluble in cold alcohol. *A solution of lime* added to a solution of citric acid or citrates causes no precipitate in the cold (differing from tartaric, oxalic acids), but on boiling a slight precipitate is formed (differing from malic acid). The precipitate of calcic tartrate is soluble in a cold solution of potash, giving on boiling a gelatinous precipitate, and again dissolves on cooling (differs from citric acid), and is dissolved by acetic acid (differs from oxalic acid).

Citric acid: TESTS.

(*1*) $CaCl_2$ in neutral solutions gives no precipitate in the cold; on boiling, precipitates $Ca_3(C_6H_5O_7)_2$, soluble in NH_4HO, but *not* soluble in KHO.

(*2*) On boiling $Ca(OH)_2$, $Ca_3(C_6H_5O_7)_2$ is precipitated.

(*3*) $AgNO_3$ in neutral solutions precipitates white flocculent $Ag_3C_6H_5O_7$.

(*4*) $Pb(C_2H_3O_2)_2$ precipitates white $Pb_3(C_6H_5O_7)_2$, soluble in NH_4HO.

(*5*) When heated to redness, irritating fumes are given off different from H_2T, also notice the $AgNO_3$ precipitate does not blacken when heated, while H_2T does.

To separate H_2T from H_3Ci, add acetic acid to make acid, if not already acid; now add potassium acetate in excess, and alcohol, when potassium hydrogen tartrate is precipitated, which can be dissolved out, while the potassium citrate and acetate are soluble in water. The citric acid may be precipitated by lead nitrate, the lead afterwards removed by H_2S.

The oxalic acid has been given before.

2. Acids precipitated by Fe_2Cl_6 and not by $CaCl_2$ — benzoic ($C_7H_6O_2$) and succinic acids ($C_4H_6O_4$).

Benzoic acid: TESTS.

(*1*) It may be sublimed in a tube in acicular needles; it has an irritating vapor, and burns with a smoky flame.

(*2*) Fe_2Cl_6 precipitates buff basic benzoate.

(*3*) When heated with sulphuric acid, it does not blacken.

NOTICE. — The gum exudes from the bark of *styrax benzoin*, a tree growing on the Malay Archipelago. It contains about 12 % of the acid. The student can make it by heating putrid cow's urine with lime, filtering, concentrating the filtrate, and precipitating the benzoic acid with an excess of HCl. The hippuric acid of the urine breaks up into benzoic and glycocine, $C_9H_9NO_3 + H_2O = C_7H_6O_2 + C_2H_5NO_2$.

Succinic acid: TESTS.

(1) Fe_2Cl_6 in neutral solutions gives a reddish brown precipitate of ferric succinate.

(2) $Pb(C_2H_3O_2)_2$ gives white a precipitate of lead succinate $(PbC_4H_4O_4)$.

(3) $AgNO_3$ gives a white precipitate of $Ag_2C_4H_4O_4$.

(4) Ammoniacal chloride of barium with alcohol gives a white precipitate in dilute solutions of barium succinate, while benzoic acid does not give a similar precipitate.

3. Acids precipitated by $AgNO_3$ in strong neutral solutions — ferrocyanic acid, $H_4Fe(CN)_6$, ferricyanic acid $H_3Fe(CN_6)$, acetic acid ($HC_2H_3O_2$), and formic acid (CH_2O_2).

Ferrocyanic acid: TESTS.

(1) $AgNO_3$ precicitates white $Ag_4Fe(CN)_6$, soluble in KCN, but not in HNO_3 or NH_4HO.

(2) Fe_2Cl_6 precipitates blue $Fe_4(Fe(CN)_6)_3$.

(3) $CuSO_4$ precipitates reddish brown $Cu_2Fe(CN)_6$.

(4) $FeSO_4$ gives a light blue precipitate, and rapidly darkens. See iron for tests.

Ferricyanic acid: TESTS.

(1) Fe_2Cl_6 gives no precipitate, a greenish brown color.

(2) $FeSO_4$ precipitates blue $Fe_3Fe_2(CN)_{12}$ (Turnbull's blue).

(3) $AgNO_3$ precipitates orange $AgFe(CN)_6$ like (1) above. See iron reactions.

10

Acetic acid : TESTS.

(*1*) Solution heated with H_2SO_4 and C_2H_5OH gives acetic ether, $C_2H_5(C_2H_3O_2)$, pleasant odor.

(*2*) Fe_2Cl_6 has a deep red color, on boiling, a light brown precipitate of basic acetate, a colorless fluid.

(*3*) $AgNO_3$ precipitates white $AgC_2H_3O_2$, soluble in NH_4HO or hot water.

Formic acid : TESTS.

(*1*) Fe_2Cl_6, blood red.

(*2*) Heated with H_2SO_4, $\overline{CO}$, H_2O, the CO may be burned, giving a blue flame.

(*3*) $Hg(NO)_3$ produces a white precipitate of $HgCHO_2$, but it soon separates into Hg and becomes gray.

(*4*) $AgNO_3$ in neutral *concentrated* solutions precipitates white $AgCHO_2$, and rapidly darkens.

To separate the acids, you must first find out what acids are in the mixture by separate tests for each acid; as, for instance, suppose you have HCl, H_2SO_4 and HNO_3, and want to separate them; (1) if you add $Ba(NO_3)_2$, you precipitate H_2SO_4 as $BaSO_4$; now filter off the precipitate. (2) Add $AgNO_3$, and you precipitate HCl as AgCl; filter as before, and you have HNO_3 left in the solution. The normal nitrates are all soluble in water, and not precipitated by bases. *The point to be remembered in separating acids* is to see that the acid combined with the reagent is the same as you have in the original solution; in other words, see that *you add no new acid* to your compound. In the above, you take $Ba(NO_3)_2$, but you have HNO_3 in the solution. You might have taken $BaCl_2$, but as $AgNO_3$ is an expensive salt, you would have to take enough to precipitate all the HCl originally in the solution, and what you added by the $BaCl_2$.

The following is the method the author used in making an analysis of cranberries. Inasmuch as it is *general*, and will apply to the analysis of any other fruit, it is given, as follows :

The fruit is pressed and filtered; if boiled, you have more coloring matter to contend with, though yielding a little higher results. Add lead acetate, $Pb(C_2H_3O_2)_2$, when the tartrate, citrate, malate, oxalate, and phosphate, if present, are precipitated as corresponding lead salts, containing, also, coloring and mucilaginous matters, etc. Pour ammonium hydrate over the precipitate. Add ammonium sulphide to the filtrate. The PbS takes with it the coloring matters. Out of the filtrate the $H_2\overline{T}$ is removed by $K\overline{A}$. To the solution is added $CaCl_2$, NH_4HO and C_2H_5OH. The precipitate will contain the $H_3\overline{C}$ and $H_2\overline{M}$. The $H_2\overline{M}$ is removed by washing with boiling $Ca(OH)_2$. Pure Ca_3C_2 remains.

By the above method we found citric, malic, tartaric and oxalic acids in cranberries; while the above statement seems simple, it is quite difficult in practice.

PART II.

CHAPTER I.

EXAMINATION OF URINE.

Urine is a secretion of the kidneys, and as generally considered, may be regarded as the product of the metamorphosis of the animal tissues, etc. The most important to the medical student are the nitrogenous constituents of the urine — urea, uric acid, etc., and the coloring matters. The daily quantity exereted is from 900 to 1,500 cubic centimeters, or from forty to sixty fluid ounces.

It may be *excessive in quantity* in persons of sedentary habits, in hard drinkers, in diseases that produce increased blood pressure, in diabetes, in hysterical paroxysms, and in the winter season of the year. It is diminished, when the other secretions are increased, as in excessive perspiration, in acute febrile diseases, dropsy and other watery discharges, and in summer, generally.

The specific gravity of normal urine varies according to food, age, sex, and the constitution of the person, and may range from 1,005 to 1,030, *and should always be based upon the entire quantity passed in twenty-four hours.* The specific gravity is *diminished* when the skin is not acting, after the free use of water and diuretics. In most chronic diseases, except diabetes, the solid residue of the urine is diminished; an increase of the solid residue indicates a more active metamorphosis, and better nutrition, and therefore is a favorable sign, as a rule. On the other hand, an increase of the solid residue of urine at the height of an acute disease is usually an unfavorable sign, because the inanition, which always occurs in such cases, is thereby increased and favored.

The specific gravity is *increased* when the urine contains sugar, urea, blood, pus, bile and albumen. The urine of females is generally of a lower specific gravity than that of males.

The *color* varies from *light* to *dark* or *black*. It is *light* colored when the quantity of urine is excessive — in anaemia, diabetes, mellitus and insipidus, in albuminaria. When the quantity is diminished, while the solids remain normal, or when an excess of animal food is used, it is dark colored. In the morning it is darker colored than at other times during the day. Mucus makes it foggy; many articles of food and medicine affect the color of the urine. Add a few drops of hydrochloric acid, and boil, and after twelve hours examine in a small beaker by transmitted light, to note if the color is permanent.

Normal urine has a characteristic *odor* while warm. The odor has no great importance to the physician; he must remember that many articles of food and medicine impart to it an odor peculiar to themselves. A *putrid* odor may be due to the decomposition of urine containing pus, blood, or albumen. An *ammonical* odor tells the story of decomposition in the system, or out of it, the urea (CH_4N_2O) becoming changed to the carbonate of ammonia, ($NH_4)_2CO_3$, as follows :

$$CH_4N_2O + 2H_2O = (NH_4)_2CO_3.$$

This change is due to microcosmic germs, which decompose the urea as yeast decomposes sugar in alcoholic fermentation.

Normal urine has an *acid* reaction, which is not due to the presence of any *free* acid(?), but to an acid salt of sodium phosphate. It is a general rule, that the urine of carniverous animals is *acid*, while that of the herbivora is alkaline, and in omnivera the reaction is intermediate, the urine being alkaline after a mixed meal. The acidity is greatest about twelve hours after taking food. The amount of acidity varies, being equivalent to from two to four grains of oxalic acid in twenty-four hours, or from .1 to .2 grains per hour. The urine may become alkaline from the blood, from ferments introduced by instruments as catheters, or after it is voided, or by decomposition of urine in the bladder, caused, it may be, by some derangement of the bladder or urinary passages.

The alkaline condition of the urine in most cases is due to the arrest of muscular metamorphosis, weakness of the nervous system, anaemia and chlorosis, defective nutrition and general debility. . Urine rendered alkaline by the carbonate of ammonia does not color litmus a *permanent* blue color, while the *fixed* alkalies do.

The chemical analysis of normal urine must vary with diet, age, sex and occupation, as well as condition in life.

COMPOSITION OF NORMAL URINE.

ORGANIC.

Water..	950.00
Urea	26.20
Creatinine	.87
Sodium and potassium urates	1.45
Sodium and . potassium hippurates	.70
Mucus and coloring matters	.38

INORGANIC.

Acid sodium phosphates	.40
Sodium and potassium phosphates	3.35
Lime and magnesium phosphates	.83
Sodium and potassium chlorides	12.55
Sodium and potassium sulphates	3.30

1,000.03

Examine urine and the sediment for blood, pus, cystine, urea and spermatozoa by the microscope.

The sediment is best examined by letting the urine stand in a conical wine glass, when the urine can be syphoned off.

A SCHEME FOR THE EXAMINATION OF URINE.

1. Odor.	5. Specific gravity.
2. Color.	6. Daily quantity.
3. Transparency.	7. Deposit and its character.
4. Reaction to test paper.	8. Confirmatory tests with reagents.

GENERAL REACTIONS.

	PRECIPITATES.	DISSOLVES.
Heat.	Albumen. Phosphates.	Urates.
Nitric acid, fifteen drops to one drachm of urine.	Albumen.	Phosphates.
Liquor potassa, KHO.	Phosphates.	Albumen.
Acetic acid, $HC_2H_3O_2$.	Cystine.	Albumen.
Silver nitrate, $AgNO_3$.	Chlorides.	
Barium chloride, $BaCl_2$.	Sulphates.	
Evaporate to one-third of its volume; add one-half the quantity of nitric acid.	Nitrate of urea.	

Excess of urea indicates............	}	Increased metamorphosis.
Excess of phosphates indicates........	}	Rachitis. Mollities ossium. Tendency to the formation of urinary calculi.
Excess of albumen indicates.......	{	Cardiac disease. Kidney disease.
Excess of specific gravity indicates..	}	Albumen, sometimes. Sugar. Urea.

The odor, color, transparency, reaction, and daily quantity have been sufficiently described, and it only remains to say a few words about specific gravity. The specific gravity can be obtained

1. By the *urinometer*, but it should be tested as to the accuracy of the scale before being used, the instrument brought to the level of the eye, and see that the spindle plays freely in the liquid.

2. By weighing a glass bulb filled with mercury just enough to sink it in the densest urine. It is on this principle that the loss of weight which the same body suffers in different fluids is proportioned to the specific gravity of the fluid. The bulb is, of course, previously weighed once for all in distilled water; an accurate balance is required in this process, and it is not much used.

3. By means of the specific gravity bottle that holds a certain given weight of distilled water under normal conditions of temperature and pressure. If now it held 1.84 times this weight when filled with sulphuric acid, the specific gravity of sulphuric acid must be 1.84, and in like manner you determine the specific gravity of any other liquid.

If it should happen that you have only a small quantity, say a table spoonful, it can be diluted with a given number of times its bulk of distilled water, say four times its volume, then there will be five volumes; if you multiply the decimal part of the specific gravity by five, you have the original specific gravity, an example will illustrate the case. The specific gravity of a sample of urine was found to be 1.024; a small portion of this was diluted with three times its bulk of distilled water carefully measured, and the specific gravity of the mixture found to be 1.006;. now this multiplied by the number of volumes, four, gives me six multiplied by four is twenty-four, or 1.024, the same as the original.

The *solid ingredients* of the urine can be approximated by multiplying the decimal of the specific gravity by two (Trapp) or 2.33 (Hæser), which will give the amount in grams of solid

contained in 1,000 cubic centimeters of the urine. We verified
Trapp's formula by four analyses; that of Hæser is too high.

Urea (CON_2H_4) is found in the urine of mammalia, birds and
reptiles, and is, in man, one of its chief constituents. It was
discovered by Boerhaave before 1720. Urea is found normally in
the blood, bile, liver, amniotic fluid, vitreous and aqueous
humors, and in the sweat; the urine contains from 2.5 to 3.5 %.
It was one of the first organic bodies that was made syntheti-
cally; it was first made by Wœhler in 1828.* It appears to
be the vehicle by which nearly all of the nitrogen of the exhausted tissues of the body is removed from the system. It is a
solid, crystalline, and colorless when pure. The nitrate is made
as follows:

The urine is filtered, and boiled to one-fourth or one-fifth of its
bulk, when nitric acid is added equal to one-half or three-fourths
of the bulk of the urine. The pasty mass is now dissolved in
boiling water, and allowed to crystallize; if now the urea be
desired, the crystals are again dissolved in hot water, and treated
with $BaCO_3$ in small portions as long as effervescence is percepta-
ble; the nitric acid has now combined with the baryta. It is now
filtered, and the clear liquid evaporated on the water bath. The
dry residue of urea and baryta is boiled with alcohol, which
dissolves only the urea and not the baryta; it may be decolorized
by animal charcoal. Or the baryta may be added first, and the
inorganic salts thrown down with the baryta, and then proceed
as before.

All tests for urea must be applied to the concentrated urine;

---

* Urea can be made artificially as follows:

A mixture of potassic ferrocyanide (fifty-six parts) and manganic peroxide
(twenty-eight parts) is heated to redness, when the following changes occur —
$K_4FeCy_6 + O_9 = 4KCyO + 2CO_2 + N_2 + FeO$. The residue is now treated with cold
water, and the clear filtrate containing the potassic cyanate (KCyO) is mixed with
ammonic sulphate (forty-one parts). The solution is then evaporated to dryness, and
the urea separated from the potassic sulphate by solution in hot alcohol — $2KCyO +
(NH_4)_2SO_4 = K_2SO_4 + 2CO(NH_2)_2$. In a general way it can be said that in the
animals that drink freely the nitrogen is excreted as urea, while in those that drink
but little it is excreted as uric acid.

if albumen is present, it must be coagulated and removed. *Urea and uric acid are tests for the presence of urine.*

(*1*) Mercuric nitrate gives a white flocculent precipitate, which may contain three or four equivalents of Hg to one of urea.

(*2*) The excessive specific gravity of the urine may point to urea or sugar.

(*3*) Nitric or oxalic acids, preferably the first, produce six sided quadrilateral plates when examined with the microscope.

The methods for quantitative analysis are quite numerous, Liebig's method of mercuric nitrate using an indicator of sodium carbonate; Knop and Huefner using a bromized soda solution, and collecting the nitrogen gas :

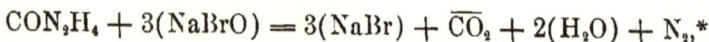

$$CON_2H_4 + 3(NaBrO) = 3(NaBr) + \overline{CO_2} + 2(H_2O) + N_2,*$$

one gram of urea yields 370 cubic centimeters of nitrogen gas. A modification of this process is used when the CO_2 is made to combine with the barium, and weighed as barium carbonate. We have not found any of these methods accurate, but like the Huefner the best; this includes all the nitrogen in the urea, uric acid, etc.†

---

*Take 100 grams of NaHO in 250 cubic centimeters of water, adding twenty-five cubic centimeters of bromine to the cold solution; fifty cubic centimeters of this solution is diluted with 200 cubic centimeters of water for use. *It is best to prepare a fresh quantity of this solution each time.*

†To obviate this trouble, it has been proposed to add sulphuric acid to make strongly acid; warm, but not boil, now titrate with a milli normal solution of potassium permanganate, which oxidizes all the nitrogen compounds except urea. The urea is now estimated by sodium hypobromite, about five cubic centimeters of the urea being taken. The uric acid is estimated by taking fifty cubic centimeters of urine, acidifying with sulphuric acid, adding baryta water in excess; this precipitates uric acid, and phosphates and sulphates; wash. Introduce in a flask, and treat with sulphuric acid, and determine uric acid by potassium permanganate solution — one cubic centimeter of permanganate is equal to 3.333 milligrams of uric acid, creatinine, or .2 milligrams of glucose and lactic acid. After the urea above is

Urea is an alkaloid of animal origin. An alkaloid is an organic base, built upon the type of ammonia, that combines with an acid to form a salt. It differs from some of the bases in this respect—it forms salts by direct union with acids without elimination of water or any other substance; in other words, by addition. Urea (CH_4N_2O) is sometimes called a *carbamide*—

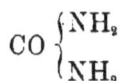

$$CO \begin{cases} NH_2 \\ NH_2 \end{cases}$$

An *amide* is a compound of amidogen (NH_2) and an acid radical. If we write carbonic acid H_2O, CO_2, or H_2CO_3 or $CO(OH)_2$ (the last is when the hydroxyl is represented), we have carbonic acid—

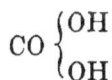

$$CO \begin{cases} OH \\ OH \end{cases}$$

If we replace the hydroxyl (OH) by amidogen (NH_2), we have carbamide, or urea—

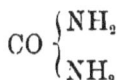

$$CO \begin{cases} NH_2 \\ NH_2 \end{cases}$$

Urea increases in quantity in diabetes and fevers, and when the diet is mostly animal food. It diminishes in quantity in cachexia, parenchymatous, nephritis, and with a vegetable diet, when fasting. The amount of urea eliminated in young children is greater per body weight than for grown persons. In old persons the elimination diminishes. Females discharge less urea than males, except during pregnancy, when it is very much increased. The average amount is .5 gram for each kilogram of body weight (six grains for every two pounds).

Uric acid ($C_5H_4N_4O_3$, $2H_2O$) was discovered and described by Scheele, 1776. It is practically insoluble in water, alcohol, and

---

estimated, take five cubic centimeters of urine, which are now treated with twenty cubic centimeters of mecuric nitrate (containing seventy-two grams of HgO to the liter), which precipitates all the nitrogenous bodies. The difference between this and the permanganate in the first operation gives that required to oxidize the nitrogenous substances ether (one part requiring 15,000 parts of cold and 18,000 parts of hot water to dissolve it), and is immediately precipitated on being freed from its bases. It is rarely in its free state in urine, and is generally combined with potash, soda, and ammonia, sometimes with lime and magnesia as *mixed urates.* The quantity is from .4 to .8 grams (six to twelve grains) in twenty-four hours. It sometimes accumulates in the vessel used to receive it, and is commonly called "*brick dust*" deposit.

Like urea, it represents metamorphosis of nitrogenized tissues, and is recognized as the first step in the formation of urea, although it is not probable that all the urea in the body passes through this stage. The acid is dibasic, and forms correspondingly two series of salts — neutral and acid. The neutral salts are readily soluble in water, the acid salts with difficulty — one part of the acid urate of sodium is soluble in 124 parts of hot and in 1,150 parts of cold water. It is found in the urine of all classes of animals, and occurs even in the very lowest orders. There is from fifty to eighty times as much urea as uric acid in the urine.

The student must remember, when testing for albumen (by pouring HNO_3 down the side of a wine glass or test tube, between the two fluids which do not mix), there appears a thick white cloud, often mistaken for albumen; but it consists of an amorphous compound that, by standing, becomes uric acid.

TESTS.

(*1*) If a little of the acid, in the state of a dry powder, is placed in a drop or two of tolerably strong HNO_3, in a watch glass or strip of glass, it will gradually dissolve — evaporate nearly to dryness at a gentle heat. *Red residue, when cold,* should be moistened by a drop or two of ammonia, or the fumes of ammonia, when a beautiful purplish-red color appears, owing to the formation of murexide ($C_8N_6H_5O_6$). Uric acid, when dissolved in dilute nitric acid, yields alloxantine ($C_8H_4N_4O_7$); and when this is treated with

ammonia as above described, yields a purple product
called murexide, as follows:

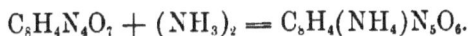

$$C_5H_4N_4O_7 + (NH_3)_2 = C_5H_4(NH_4)N_5O_6.$$

If you now add a little potash, a beautiful purple color
is produced.   Potash is better than ammonia in the
above.   All urates give this test:

(2) When heated before the blowpipe, a disagreeable smell,
resembling that of burnt feathers mixed with hydrocyanic acid, is given off.   This shows it to be an
organic nitrogenous body.

(3) By the microscope the peculiar "whetstone" shaped
crystals, yellowish-brown color, *white when pure.*

(4) Dissolve the uric acid in a solution of sodium or
potassium carbonate, place a few drops on paper;
now add a solution of nitrate of silver.   A gray stain
indicates uric acid.   Uric acid is soluble in an alkali
at a high temperature, and in sulphuric acid, without
decomposition; for it can be re-precipitated by the
addition of water.

(5) Add to a solution of uric acid in potassic hydrate an
alkaline copper solution; a white precipitate of
cuprous urate is formed.   If, now, this be heated to
boiling with an excess of the solution of copper, the
uric acid oxidizes, and red cuprous oxide separates,
while the oxidation products of uric acid dissolve.

(5) In testing the blood-serum of gouty patients for uric
acid, when only a small quantity can be obtained:
For every four to eight grams add six to twelve drops
of strong acetic acid, and after, a thread about an inch
long is laid in the fluid for eighteen to twenty-four
hours, at a temperature of 16° to 20° C.   The thread
can then be examined with the microscope for uric
acid.   For the oxidation products of uric acid see
Tidy's chemistry, page 716.   The urates act like uric

acid: (1) They are soluble in an alkali; (2) Acid
decomposes them, and liberates uric acid — examine
under the microscope; (3) They disappear at 100° F.,
when heated.

PREPARATION.

Filter the urine, and add twenty cubic centimeters of HCl
to one liter of urine; let stand in a cool place twenty-four to
forty-eight hours, filter and test. They act like sand under a
glass rod, or feel quite gritty.

This method has two sources of error: (1) A certain amount
of uric acid always remains in solution; (2) It brings down
coloring matter with it.

Another method is to throw down the urine with acetate of
mercury; now decompose this with sulphuretted hydrogen, and
determine the uric acid in the filtrate.

Thudichum gives this as the best method, after trying many
experiments: "The urine is evaporated to dryness, treated with
four or five times its volume of strong alcohol (90%), well
shaken, and allowed to stand for twenty-four hours, the fluid
filtered off and the deposit washed with alcohol until colorless.
The deposit is treated with water and hydrochloric acid for phosphates. The residue of uric acid is washed and dissolved in
caustic potash, warmed until all ammonia is gone, and reprecipitated by the cautious addition of hydrochloric acid until the
fluid, which should be still warm, has an acid reaction. It is
allowed to stand in a cool place for twenty-four hours, when the
uric acid may be isolated and weighed in the usual manner."

Uric acid is *increased* by rich diet, animal and vegetable, and
too little exercise in open air; in lung and heart diseases, with
dispnœa; in acute febrile diseases, which cause much breaking
down of the nitrogenous elements of the body; in leucæmia —
impoverished blood; in case of large tumors of the abdomen,
ascites, etc.

Uric acid is *decreased* in chronic affections of the kidneys,
diabetes mellitus (occasionally), urina spastica, hydruria, and
arthritis.

Albumen $(C_{72}H_{118}N_{18}O_{22}S(?))$.—The connection of certain forms of dropsy with the presence of coaguable albumen in the urine, and of the latter with kidney disease, was first pointed out by Bright, 1827. Albumen is formed in plants, and finds its way into the animal system as food, when it passes through the many complex changes and finds its exit from the system as *urea*. But when albumen as such is discharged in the urine, it indicates one of two things—either severe disease of the blood or of the kidneys. From this it can be seen that it is found in *abnormal* urine. In a general way, most *mineral* acids precipitate albumen (except tribasic phosphoric acid), while most *organic* acids do not precipitate it. A solution of albumen causes a turning of the plane of polarization towards the left. The albumen of the blood is not precipitated by dilute sulphuric acid, yields no hydrosulphuric acid on coagulation by heat, and contains less sulphur than the albumen of eggs. In most other respects they are identical.

Albumen is the "ornithorhynchus" of the secretions, for it is found in acid, neutral, and alkaline, solutions of the urine, in high colored and in pale, in high specific gravity and in low. Albumen is coagulated by heat, alcohol, and the mineral acids.

TESTS.

(*1*) Heat coagulates albumen. Now add a few drops of nitric acid; this dissolves the earthy phosphates, and a white precipitate indicates albumen. This is a very delicate test. The test tube containing the urine should be held obliquely, and the pure colorless nitric acid added by a pipette in drops; a sharp *white zone* indicates albumen, with a brown line of demarcation between the two.

(*2*) The microscope can confirm, if there is any doubt as to albumen, or urates, or uric acid.

(*3*) A rough test is usually made in the sick room by heating the urine in some vessel; as, a spoon, and noticing the result.

(*4*) Treat the urine with a few drops of acetic acid
(H, C₂H₃O₂) to make it have a strong acid reaction;
now add a few drops of ferrocyanide of potassium
(K_4FeCy_6), a white flocculent precipitate indicates
albumen ; this is a very delicate test.

(*5*) If acetic acid be added as in the above, and it be treated
with an equal volume of a saturated solution of
sodium sulphate (Na_2SO_4), and then boiled, complete
coagulation results ; this is quite a good test.

NOTICE. — *Pure* albumen is not coagulated by boiling temperature, or by alcohol; also, sugar and sulphuric acid produce a
beautiful red color with albumen and all protein bodies just as
with the biliary acids. Millon's reagent is made as follows :

Mercury is dissolved in its own weight of strong nitric acid ;
this is now diluted with twice its volume of water, which gives a
white precipitate with albumen.

Next to urea, the chlorides are the most abundant in the urine,
for they are taken with almost every article of food that we use.

The chlorides are chiefly those of sodium, with a small proportion of potassium and ammonium. If Liebig's method is
used, the phosphoric and sulphuric acids are first removed by
the addition of the nitrate and hydrate of barium. It is made
acid by the addition of nitric acid, and it can now be determined
by a solution of mercuric nitrate. The precipitate that forms is
a compound of urea and mercuric oxide; in the presence of
common salt, the mercuric nitrate is changed into corrosive
sublimate, which does not precipitate the urea in slightly acid
solutions. The first precipitate that forms is redissolved until a
permanent precipitate is formed. If Mohr's method is used, you
take fifteen or twenty cubic centimeters of the urine in a
platinum crucible to which are added two or three grams of
potassium nitrate, free from chlorine, and evaporate to dryness
on the water bath; this is now ignited, at first gently, then
strongly, to destroy the organic matter, the ash being white;
dissolve in water; add one or two drops of nitric acid to make it

slightly acid; this is *exactly* neutralized by calcium carbonate, and without filtering, a few drops of a neutral solution of chromate of potash is added. Silver nitrate is now run in until the red chromate of silver begins to form. If bromine or iodine have been taken as medicine, the residue must be treated with potassic nitrite, a few drops, to separate them, and they are then taken up with bisulphide of carbon (CS_2); it is neutralized with sodium carbonate, and treated as before.

If the albumen is present in large quantities, it should be removed before using these tests; but if only in small quantities, it can be omitted. The simple test is to add a few drops of nitric acid and silver nitrate, when a white curdy precipitate indicates chlorides.

Phosphates exist in the urine in two forms—*alkaline* and *earthy;* the alkaline are soluble, and are never met with as deposits in the urine, while the earthy are commonly as precipitates. The phosphates are sometimes mistaken for albumen, for the reason, that they are precipitated on heating the urine; but they are soluble in nitric acid, while albumen is precipitated. The earthy and alkaline may be approximately separated as follows:

Add ammonia in excess, and boil it; earthy phosphates are precipitated; filter, and add "magnesium mixture"; a *white crystalline* precipitate indicates alkaline phosphates.

For the tests for phosphoric acid see the bases, page 41. A rough practical test of the quantity, for the physician, can be had by taking a test tube containing three or four inches of filtered urine; now add an alkali as KHO or NH_4HO, a few drops, and warm gently over a lamp until the phosphates begin to settle out. If in normal quantity, they will occupy one-third of an inch in height in the test tube.

TESTS.

(*1*) Silver nitrate throws down a *yellow* precipitate of phosphate of silver, and also a chloride of silver. The chloride is soluble in ammonia, as well as the phos-

12

phate; but nitric acid dissolves the phosphate, but not the chloride; they can be separated in this way.

(*2*) Ammonium molybdate in the presence of nitric acid, when heated, gives yellow precipitate if only traces are present.

(*3*) If a hot solution of a phosphate that has been dissolved in water or acetic acid be now treated with a solution acetate or nitrate of uranium, when a phosphate of uranium falls, whitish yellow.

Grape Sugar ($C_6H_{12}O_6$) has been known to exist in the urine for nearly a century (1792). It can be made artificially from cane sugar and starch by boiling with dilute acids. It turns the plane of polarisation to the right. It is less soluble than cane sugar. It does not blacken with sulphuric acid, like cane sugar. It is a disputed question whether sugar belongs to normal or abnormal urine. Sugar might, and in most cases is, found in the urine of mothers after their weaning of children. The diseases in which sugar occurs have been sufficiently indicated before. The urine should be filtered, and the albumen, if present, removed by nitric or acetic acid. It is now decolorized with animal charcoal, and fifteen or twenty cubic centimeters taken and diluted with three or four times its volume of water, and the following tests applied. A word might be said, in a general way, about tests. Much of the difference of opinion with regard to the reliability of the different tests is due to the fact that those claiming great things for them have had more experience with the particular test which they recommend than with any other, or perhaps all others combined. It is well to select some one test and make it your own, but you should not be dependent upon *any one test.*

TESTS.

(*1*) Moore's. — Take a test tube half full of urine, and add one-half as much caustic soda or caustic potash, and boil; yellowish-brown color indicates sugar. The earthy phosphates can be filtered off before the test

is applied. If they are very abundant Heller's modification is to treat the fluid with nitric acid, when
the odor of burnt molasses is given off.

(2) Fehling's and Pavy's are about alike. The solutions are
made as follows* :

PAVY'S SOLUTION.	FEHLING'S SOLUTION.
Cupric sulphate .... 320 grains.	Copper sulphate .... 34 grams.
Neutral potassic tartrate 640 "	Rochelle salts .... 180 "
Caustic potash. 1,280 "	Solution caustic soda... 600 "
Distilled water......... 20 fl'd oz.	Sp. gr., 1.12.
	The whole diluted to a litre.

The copper sulphate is dissolved in the water, the tartaric acid
and caustic soda mixed, and this mixture added to the copper
solution.

The solution should have a clear blue color. The solution
should be boiled, and remain clear. If sugar is present, the red
suboxide (Cu_2O) is thrown down on heating. The best way is to
mix these dry, and use when required, as follows :

1 part of $CuSO_4$ .... Copper sulphate.
5 parts of $KNaC_4H_4O_6$ .... Rochelle salt.
2 parts of NaHO .... Caustic soda.

Take a piece about the size of a kernel of corn, and dissolve it
when required, or keep the solutions in separate bottles, and mix
them as you want to use them.

(3) Boettger's bismuth test.—Take equal quantities of urine
and caustic potash or soda, and add a piece of the
subnitrate of bismuth the size of a large pea, and boil
for a couple of minutes; if sugar is present, black
metallic bismuth is precipitated. Sulphur will cause

---

* It must be remembered that uric acid will reduce a Fehling solution, as well as
grape sugar. There is a great number of substances that will reduce it; but they would
not be found in the urine, as aldehyde, chloral, arsenious acid, etc.

the same reaction; if the urine contains sulphur, add a drop or two of hydrochloric acid, and treat with a solution of iodide of potassium and bismuth; this removes the sulphur, but does not affect the sugar. The solution can be filtered, and you can now use Moore's test (1). It can be seen from this that all the albumen should be removed, as, by the formula, it contains sulphur.

(*4*) Mulder's test. — The urine is heated with a solution of indigo, carmine, which has been previously made alkaline by sodium carbonate. The blue mixture becomes green, then purple red, and finally yellow.

(*5*) The fermentation test. — Twenty or thirty cubic centimeters of the urine are placed in a suitable flask, and a little dry yeast and a small amount of tartaric acid is added; the object of the tartaric acid is to prevent other decompositions, while, at the same time, vinous fermentation is promoted. The carbonic acid formed is dried by passing it through sulphuric acid, and caught in potash bulbs, and weighed. Pasteur has shown that other substances than carbonic acid result from the fermentation of sugar; as, amyl alcohol ($C_5H_{12}O$), butyl alcohol ($C_4H_{10}O$), succinic acid ($H_2C_4H_4O_4$), and glycerine ($C_3H_8O_3$).

Sulphates are found with the alkalies, potassium and sodium, partly with lime as gypsum. Add to the clear urine a few drops of hydrochloric acid to keep up the phosphates; now add $BaCl_2$, when a white insoluble $BaSO_4$ is precipitated. The amount in normal urine gives a milky appearance. The amount of sulphuric acid found is, in a measure, an index to the metamorphosis of the muscular tissues. Urine containing *mucus* is generally cloudy, ropy, and alkaline. If the urine contains *pus*, it always contains albumen; the pus settles to the bottom readily. If it contains *fat*, it is milky, opaque, and albuminous; the fat comes to the surface readily on standing.

Albumen and sugar are rarely present in the same urine. The same can be said of the sediments of the phosphates and urates. In all the tests with urine, the abnormal ingredient can be added to it — for albumen, the white of egg, etc.

The coloring matters of the urine are divided into two classes — normal, urochrome and indician, and abnormal, blood, bile, vegetable coloring matters, and uroerythrin. If we except the blood, which will be treated of in another place, and the bile, the rest are not of sufficient importance to demand our attention.

Bile is a secretion of the liver, with a color from green to black. It has a specific gravity of 1.020 to 1.030; taste, intensely bitter, and reaction, alkaline to test paper. Sodium glycocholate and taurocholate are the principal bile salts. Bilirubin and biliverdin are the chief bile pigments; a little ox gall may be added to the urine for the purpose of applying the tests. When biliary matter is present in urine, the color is more or less brownish yellow.

TESTS.

(1) Remove the albumen, if present; next add strong sulphuric acid and sugar. A play of colors from pink to purple indicates bile salts; this is known as *Pettenkofer's test*. In practice, the urine is concentrated nearly to dryness on the water bath, the residue treated with boiling water or alcohol, and treated as before described.

(2) *Heller's* modification is to mix the urine with albumen, and shake them well together; now precipitate the albumen by nitric acid, and, if bile is present, the precipitate will be more or less dull green or bluish color.

(3) *Gmelin's* test for pigments. — Take a few drops of the solution, as above described, on a porcelain plate; add a few drops of yellow nitric acid; if bile is present, the liquid becomes successively pale green, violet, pink, and yellow, the colors rapidly changing as the urine is mixed with the acid.

CHAPTER II.

GENERAL METHOD FOR THE TREATMENT OF URINE.

The reaction is ascertained with litmus paper. *If neutral or alkaline*, a sediment is usually formed, and should be examined with a microscope; *if acid*, it usually forms no sediment, but if a sediment is formed, it should always be examined.

1. The urine is filtered, and, if necessary, made slightly acid with a *few* drops of acetic acid ($HC_2H_3O_2$), and boiled; add to the precipitate, nitric acid (HNO_3); if it does not dissolve, it indicates *albumen*. When the precipitate is white, it indicates pure albumen.* When the precipitate is greenish, it indicates biliary matters.† When the precipitate is brownish red, it indicates blood.‡

2. Six hundred cubic centimeters of the filtrate or original solution are evaporated to a thick syrup on the water bath, and divided into two portions — (*a*), one-third, and (*b*), two-thirds. (*a*) This portion is treated with strong alcohol three or four times, and the *solution* and *residue* separately examined. The *solution* is divided into two portions — one-fourth and three-fourths. The one-fourth is taken and evaporated nearly to dryness on the water bath, and the residue tested for *urea*. See page 84. The three-fourths of the solution is treated with milk of lime, and then with a concentrated calcium chloride as long

---

*Confirm by taking equal parts of carbolic and acetic acids, with two parts of 90 per cent. alcohol; now two or three parts of nitric acid and ten parts of this solution are added to the urine, and the mixture is shaken and allowed to settle: a deposit indicates albumen. The deposit is hastened, if a saturated solution of sodium sulphate, in the place of nitric acid, is used. It can also be confirmed by the other tests given for albumen.

† Confirm by tests on page 95.

‡ Confirm by tests for blood.

as a precipitate is produced. The filtrate is evaporated on the water bath to ten or twelve cubic centimeters, and one-half of a cubic centimeter of chloride of zinc in alcohol is added, and the whole well shaken, when *kreatinin-chloride of zinc*, white. Test with the microscope. (*b*) This portion, two-thirds of the residue, is slightly acidulated with hydrochloric acid, and well mixed with powdered sulphate of barium, and extracted with alcohol; this solution is tested for hippuric acid.

3. The *residue* (1) is tested with dilute hydrochloric acid (one part to six), and filtered. The solution contains the *earthy phosphates* and other salts; the phosphates are precipitated by neutralizing with ammonia. The residue is *mucus* and uric acid; if this be heated with sodic hydrate, the uric acid dissolves in the filtrate, and leaves the mucus behind. These can now be examined by their respective tests, see before.

The ingredients of the urine, when large quantities of it are required, are omitted, and only those are dwelt upon, that are of practical importance to the physician. The microscope is invaluable in the examination and in the identification of pus, mucus, blood, most of the salts, fat, epithelium casts, spermatozoa, and parasites. These are figured in the larger works on the analysis of urine, and can be compared and consulted by the student. A peculiar deposit of membrane which forms on the surface of long standing urine of pregnant women, to which the name of *kyestein* has been given, it is now known that it forms upon the urine of men, and was once considered a characteristic sign of pregnancy, but is now generally discarded.

Nearly every article of medicine can be detected in the urine, when given in excess to the patient. The kidneys secrete from the blood, bodies in an unaltered state (As, Sb, Bi, Cu, Cr, Au, Fe, Li, Pb, Ag, Sn, Zn), and free organic acids and many of the alkaloids; as, morphine, strychnine, and atropin, besides many oxidized or changed products. The composition of the urine varies hour by hour. The morning urine (*urina sanguinis*) consists chiefly of the products of tissue metamorphosis, and is usually taken for analysis, while the day and evening urine (*urina cibi and urina potus*) is influenced by the quantity, and also by the character of the food ingested.

QUALITATIVE ANALYSIS OF URINRY CALCULI.

Powder the calculus; heat a small portion on a platinum foil, noticing if there is a residue.

1. If there is a residue, take a small portion of the original, and dissolve it in strong nitric acid, and evaporate to dryness on the water bath in a porcelain dish; bring a drop of the strongest ammonia near the residue in the dish, and observe whether a *pink* color is produced or not. (*a*) If *pink color*, it contains *uric acid*.

(*1*) It melts and colors the flame of the lamp; when *yellow*, sodium urate; when *violet*, potassium urate.

(*2*) It does not melt. Dissolve the residue after ignition in a little hydrochloric acid, and add ammonium hydrate until alkaline, and then ammonium carbonate; if a white precipitate, indicates calcium urate; if there is no precipitate, then add Na_2HPO_4, when a white crystalline precipitate indicates Mg urate.

(*b*) *No pink color.*—Heat a portion, and notice if it melts.

(*1*) When it melts (fusible calculus), treat the residue with $HC_2H_3O_2$ when it dissolves; add NH_4HO in excess; a white crystalline precipitate indicates ammonio-magnesium phosphate. If the melted residue is insoluble in $HC_2H_3O_2$, treat with HCl, it dissolves, now add NH_4HO, white precipitate $Ca_3(PO_4)_2$.

(*2*) It does not melt; moisten residue with water, and test the reaction with litmus paper; it is not alkaline. Treat with HCl, dissolves without effervescence; add NH_4HO in excess, white precipitate $Ca_3(PO_4)_2$. Treat the calculus with $HC_2H_3O_2$, it does not dissolve; treat the residue after treating with $HC_2H_3O_2$; it dissolves with effervescence $CaCO_3$.

2. The calculus on being heated does not leave a fixed residue. Treat a portion with HNO_3; evaporate, and expose as before to NH_4HO vapor. (*a*) If a *pink color*, mix a little of the powdered calculus with CaO, and moisten with water, and heat $\overline{NH_3}$; test by litmus paper and HCl, indicates *ammonium urate*. (*b*) No NH_4HO indicates uric acid.

3. No pink color. (*a*) HNO_3 solution turns yellow as it is evaporated, and leaves a residue, insoluble in potassium carbonate, and indicates *xanthine*. (*b*) HNO_3 solution turns dark brown, and leaves a residue, soluble in ammonia, and indicates *cystine*.

PART III.

CHAPTER   I.

EXAMINATION OF WATER.

Water (H_2O) was regarded by the ancients as an element. Priestly, in 1780, correctly predicted its composition. It is one of the most abundant compounds in nature, but never found pure. It requires about 2,600 volumes of the gases (H and O) to make one of the liquid. It has neither color (when viewed with transmitted light, blue), odor, or taste when pure. Its specific gravity is taken as the standard of comparison for liquids and solids. It may exist as a solid, liquid, and gas. It is eight hundred and twenty-five times heavier than air; a cubic foot weighs 1,000 ounces (997 actually); one gallon, 7,000 grains; one liter, 1,000 grams, or a kilogram. The specific gravity of ice, .9184: Water in the act of freezing expands — 174 volumes of water at 60° F. becomes 184 volumes of ice*; one volume of water at 0° C. becomes 1.09082 volumes of ice. It is practically incompressible. Every atmosphere contracts it about fifty-one millionths of its bulk. The latent heat of water is (143° F.) 79° C.; the latent heat of steam, 537° C., or 537 thermal units. One cubic inch of water will make nearly one cubic foot of steam (1,696 cubic inches).

Water has its greatest density at about 4° C. It is nature's universal solvent.

---

* Bismuth, cast-iron, and antimony expand on becoming solid, as well as water.

THE BOILING POINT OF WATER AT DIFFERENT PRESSURES.

Boiling point. Degrees F.	Pressure in atmospheres.
212	1
249.5	2
273.3	3
291.2	4
306	5
318.2	6
329.6	7
339.5	8
348.4	9
356.6	10
415.4	20

An elevation of 590 feet lowers the boiling point $1°$ F., 1,062 feet for $1°$ C. Many elements decompose water, taking its oxygen and liberating hydrogen, as $Na_2 + H_2O = Na_2O + \overline{H_2}$. It is an indifferent body. It may be a base, as H_2O, SO_3, or an acid, Na_2O, $H_2O = (NaHO)_2$. (1) It unites with *anhydrides* to form acids, H_2O,SO_3; (2) It unites with *bases* to form hydrates, Na_2O, $H_2O = (Na\,HO)_2$; (3) It unites with certain bodies as water of crystallization; this is more or less connected with (a) the shape (b) the color of the crystal. Thus, if blue sulphate of copper looses four molecules of water, it is *white*, and looses both shape and color. This is, in fact, the *best test* for water. It is yet more strikingly shown in magnesic-platino-cyanide, which is red, green, yellow, or white, according to the quantity of water it contains; (4) It unites with bodies forming a part of the chemical properties of bodies, and is called water of constitution; (5) This constitutional water may often be replaced by another salt, as follows: $MgSO_4$, H_2O, 6Aq. Now K_2SO_4 can replace the H_2O, as in this case, $MgSO_4$, K_2SO_4, 6Aq (Aq represents the water of crystallization). When a body looses its water of crystallization, it is called efflorescent; but if it takes on water, it is called deliquescent.

Water composes four-fifths of our flesh and blood. A man is nothing but six pails of water combined with twelve to fourteen pounds of solid matter. Cabbage contains 92 per cent. of water; cucumbers, 97 per cent.; and watermelons, 98 per cent. Prof. Agassiz describes an entire order (acalephs) that is composed of only ten parts in a thousand of solid matter. Every gallon of water contains from seven to eight cubic inches of air. The air dissolved in water is richer in oxygen (33 per cent.) than ordinary air, which contains 21 per cent.

Magnesium rocks allow			20 per cent. of the water to pass through.					
Silicious	"	"	25	"	"	"	"	"	"
Chalk	"	"	42	"	"	. "	"	"	"
Loose sand	"	"	90 to 96	"	"	"	"	"	"

A well will drain the surface from sixty to eighty feet from it. Lefort found impurities 330 feet from a graveyard; it is needless to add that such a dilute solution of decaying humanity must be very dangerous. The impurities of water are, mainly, inorganic substances dissolved in it; as, lime, etc., and organic and other impurities held in suspension. The medicinal qualities of water are mainly due to the former, while the deleterious qualities are due to the latter, and may be a gas, a liquid or solid, or living organisms; sometimes the *odor* betrays the presence of one or more of these impurities. The ordinary water analysis includes hardness, poisonous metals, chlorine, ammonia (free and albuminoid), and total solids. The quantity of water usually taken is seventy cubic centimeters. An English gallon * of water weighs 70,000 grains, and seventy cubic centimeters of water weigh 70,000 milligrams; so seventy cubic centimeters of water is a *miniature gallon* where the milligrams correspond to grains. This gallon is taken and evaporated to *dryness* on the water bath in a platinum dish; the increased weight is the *total* solids. This is now heated over a Bunsen burner until white; the loss is *organic* matter, and in some cases a *little acid* as carbonic acid. This

---

* An English gallon is equal to about 1.2 American gallons.

residue is treated with HCl, and the poisonous metals determined by the usual tests. See qualitative separation of the bases.

Chlorine is usually in combination with sodium as NaCl. An excess (pure water contains hardly any) indicates contamination with sewage. Dissolve 4.79 grams of pure silver nitrate in one liter of distilled water; now, one cubic centimeter of this solution precipitates one milligram of chlorine. To our gallon, a few drops of a neutral solution of pure chromate of potash are added, and then the silver nitrate is run in from a burette; the excess of silver nitrate is indicated by the red silver chromate. The water should be neutral; if acid, it dissolves the silver chromate. It is a good plan to make a blank trial with distilled water *in all your experiments.*

The organic matter is the most important part of a water analysis. As usually performed, the following solutions are required.

1. Nessler's reagent is made by taking thirty-five grams of KI, and thirteen grams of HgCl₂, and 800 cubic centimeters of water. The materials are heated to boiling, and stirred up until the salts dissolve; to this is added a cold solution of HgCl₂ in water, until the red precipitate just begins to be permanent; to this is added 160 grams of KHO, or about 120 grams of NaHO, whichever is most convenient, and the whole made up to a liter; it is now allowed to settle, and then put away in a large bottle, well corked. The solution is poured into a small flask, as required for use; two cubic centimeters are used for a test.

2. For permanganate solutions, take 200 grams of potash, and eight grams of permanganate of potash in a liter of water; boil for a few minutes to get rid of any traces of ammonia; use fifty cubic centimeters for each analysis.

3. For ammonia solutions, take 3.15 grams of chloride of ammonia in a liter of water; one cubic centimeter contains one milligram. The solution can be diluted, if required. The retorts

and other apparatus, it is not necessary to consider.* It is important to notice that the distilled water *must always be examined for ammonia;* we never examined a specimen that did not contain it.

All the apparatus must be examined to see that they are perfectly clean, and, if possible, kept *only for this purpose.* Half a liter of the water is placed in the retort, and when the first fifty cubic centimeters come over, it should be Nesslerized by adding to it two cubic centimeters of Nessler's reagent, and comparing it with the standard solution of ammonium chloride by its color in two glass vessels of the same size. It has been found that this fifty cubic centimeters invariably contains three-fourths of all the *free ammonia.* To the result then add one-third of the amount found. While this fifty cubic centimeters of the distillate is tested, it should be allowed to proceed until, in all, two hundred cubic centimeters have distilled over, leaving three hundred cubic centimeters in the retort. Now, to this is added fifty cubic centimeters of the solution of potash and permanganate of potash. The object of the potash is to liberate the albuminoid ammonia; the distillation is continued until fifty cubic centimeters have come over, when it is Nesslerized. Fifty cubic centimeters more are treated in the same way, and again another fifty cubic centimeters, when the three results are added up. The Nesslerizing is comparing the color of the distillate with color obtained by a known quantity of ammonia from the ammonia chloride solution. The *total* ammonia is found by taking the free ammonia and adding one-third to it, and getting the sum of the three albumi-

---

* *A glass retort, holding about two liters, is used, having an opening at the top to add the water so as not to dislodge it (it can be cleansed by a syphon); this is fastened to a receiver of the same size by steam tight fittings, and cooled by water; or, when a condenser is used, it is caught in glass jars of uniform size holding about 100 cubic centimeters; the retort is heated by a sand bath. The buretts and other instruments are the same as for other purposes in the laboratory, and only require extra care in cleaning. It is very convenient to have pipetts of the following sizes: one, two, five, ten, twenty and fifty cubic centimeters. These can be made by the student, if he has not got them. They are very handy to have for many purposes besides water analysis.

noid* ammonias together (sometimes a fourth and fifth time, but usually the third fifty, cubic centimeters will contain scarcely any). Now, as this is for five hundred cubic centimeters, double this will be for a liter, expressed in milligrams, or what is the same thing, parts per million.

It must be noticed in regard to organic analysis that chemistry is yet in its infancy, especially with that branch that deals with the secretions of plants and animals. It is almost a rule in biology that the lower the form of life, the more tenacious it is of it, and heat and cold does not seem to affect them. Dr. Saunderson found germs of bacteria from the heart of ice, clear as crystal, in an Alpine stream 7,000 feet above the level of the sea. The chemist, the microscopist, and the biologist, being, in many cases, powerless to detect those living germs, the chemist can only say ammonia, nitrates, or nitrites, are proofs of previous organic impurities, but they all fail to say what plants, animals, or compounds impregnate it, whether these impurities are recent or remote, or if it contains those small quantities of morbific

---

*The so called wet combustion is made as follows: A solution of ferrous ammonium sulphate $(FeSO_4(NH_4)_2SO_4 + 6H_2O$·, is made so that one cubic centimeter will require one milligram of oxygen $(2FeO + O = Fe_2O_3$, or 112 of Fe is oxidized by sixteen of oxygen —seven to one); now as the above salt is one-seventh iron, you require forty-nine grams to the liter: also, make a solution of permanganate of potash so that one cubic centimeter of it will contain one milligram of oxygen,

$$2KMnO_4 + 3H_2SO_4 = K_2SO_4 + 2MnSO_4 + 3H_2O + O_5.$$

The sum of the atomic weights (316.274) gives a normal solution when divided by ten or 31.62 grams, but this (316.274) liberates five parts of oxygen (five times sixteen is equal to eighty parts by weight, 316.274 divided by eighty is equal to 3.9534). Therefore 3.9534 grams of permanganate of potash are required to a liter. In the solutions, a cubic centimeter of one will exactly oxidize the other, besides these solutions of caustic potash, and one containing sulphuric acid, are used. The retort is cleaned and mounted as for the ammonia process. Take one liter; add five cubic centimeters of NaHO and five cubic centimeters of permanganate of potash, and distil until 900 cubic centimeters have come over; now add ten cubic centimeters of sulphuric acid; shake it up, and add five cubic centimeters of the iron solution (protosulphate), when the liquid becomes colorless. The solution of permanganate of potash is now run in until a red color just begins to be permanent. The difference between the iron solution and the permanganate gives the amount consumed by the organic matter. The first quality of water requires only .5 milligrams of oxygen per liter; average drinking water, two to three milligrams: all above this is regarded as very bad water.

matter that contain the seeds of death to those that drink it, all, alike, are unable to state.

The nitrogen of the ammonia (NH_3) can come from the urea (CON_2H_4) of the urine, and shows contamination of sewage.

The sulphates are determined by Barium chloride ($BaCl_2$).

The nitrates are determined by aluminium foil; a piece larger than is required is added, and the nitric acid is converted to ammonia, as follows: Caustic soda, free from nitrogen, is added to the solution to make it strongly alkaline; a piece of aluminium foil is added, larger than it is capable of dissolving, and it is left there for three or four hours. The amount of ammonia may be determined by Nessler's reagent, NH_3 is to HNO_3 as 17 is to 63.

The *nitrites* can be detected by concentrating the water, adding a few drops of dilute sulphuric acid, a weak solution of potassic iodide, and some starch paste. The nitrites liberate the iodine; and this colors the starch blue, while nitrates do not act in this manner. One of the *most delicate* tests for nitrites is *metadiamido-benzol hydrochlorate*, a brownish colored salt, and gives, with a nitrite, a blood red color. *A good many of the articles sold for the above are not good, and will not give the above result.**

THE HARDNESS OF WATER.

Dr. Clark's method of determining the hardness of water:

1. The standard solution of soap can be made by taking ten grams of good castile soap, and dissolving it in a liter of dilute alcohol (35 per cent.). If there is any doubt about the purity of the soap, it can be compared with a solution containing a known quantity of carbonate of lime, or of chloride of calcium equal to a known quantity of carbonate of lime, as follows: 1.11 grams of pure fused calcium chloride, dissolved in a liter of water, is equal to one milligram of carbonate of lime in one cubic centimeter. Now the soap solution can be verified with this, and the correction made, if it is stronger or weaker. We take our

*See note on page 108.

miniature gallon, seventy cubic centimeters, and dilute this with once, twice, etc., times its volume of distilled water, and then add our soap solution, and shake after each addition until a lather forms, when the vessel is placed upon its side that is permanent for five minutes. It is diluted for this reason, that too large a proportion of insoluble lime salts interferes with the lathering. An allowance of one degree of hardness must be made for each seventy cubic centimeters of distilled water added. This is called the *total* hardness. If water, charged with the carbonate of lime, be boiled, the excess of carbonic acid escapes, and there is a deposit of more or less carbonate of lime, and the · water becomes softer. The hardness of the water after this deposit is called *permanent* hardness, and the difference between the *total* hardness and the *permanent* hardness is called *temporary* hardness. If the water contains magnesia as well as lime, there is not much dependence to be placed upon the result. (1) Magnesia is not instantly precipitated, and (2) it requires about one and one-half times as much soap as lime. The better way is to evaporate the water to dryness, and make an analysis of the residue in the usual way. It might be well to state that the season of the year, spring or summer, has a good deal to do with the result of the analysis. In times of high water you can expect more organic matter; also, different rocks may be washed by the water in different stages of high water or flood. Springs and deep wells afford the best water.

Organic matters may be gotten rid of (1) by boiling, (2) by the admixture of air, as water flowing over rapids, etc., (3) by filtering through carbon, alum is sometimes used, (4) by permanganate of potash ($K_2Mn_2O_8$), as follows: add one-sixth of the volume of sulphuric acid (H_2SO_4), and then a solution of the permanganate; if organic matter is present. it will decolorize it. Of the poisonous metals, lead is the most important. To test the presence of lead, evaporate the water to one-fourth of its volume, add a *few* drops of hydrochloric acid (HCl) to just · make slightly acid; now pass sulphuretted hydrogen (H_2S)

14

through; if lead is present, it will blacken or darken in color, depending upon the amount of lead present.*

Bad water may effect the stomach,, and cause dyspepsia, diarrhœa, typhoid fever, cholera, yellow fever, etc.; sulphuretted hydrogen (H_2S) may cause skin disease, and sulphurous acid (SO_2), disease of the bones; lime ($CaCO_3$) may cause calculi and goitre (see Gamgee's Physiological Chemistry).

The physician can easily see the importance of water, and a knowledge of its composition. In one case it may carry health and happiness to his patient, in the other, death and desolation. How doubly enchanting does it become, when he views it in the broader light of science; when it takes to itself the wings of

---

* The sulphide may not be lead. It can be dissolved in HCl, and tested for other bases by the separation of bases that have gone before.

NOTE.—1. Estimating nitrites in the presence of nitrates by the oxidation of gallic acid ($C_7H_8O_5$).

Gallic acid + nitrous acid = tanno melanic acid + carbonic acid + nitric oxide + water

$$C_7H_6O_5 + 2HNO_2 = C_6H_4O_3 + CO_2 + 2NO + 2H_2O$$

Twenty-five cubic centimeters of the water are taken. If iron is present, it must be removed by ammonia. To this is added one or two cubic centimeters of gallic acid solution, heated in a test tube with a few drops of hydrochloric or sulphuric acid; when it has cooled, the color is compared with standard acid and a nitrite solution of known strength, under like conditions. The gallic acid solution is a saturated solution decolorized by charcoal. The addition of HCl or H_2SO_4 prevents the change of color which may. take place after keeping. The alkaline nitrate solution is made by decomposing silver nitrite by sodium chloride; .406 grams of silver nitrite is decomposed, and the solution made up to one liter; 100 cubic centimeters of this is diluted to 1,000 cubic centimeters, when one cubic centimeter equals .01 gram of Na_2O_3. One part of acid in 20,000,000 parts of water can be detected, giving a brownish-red color. The presence of organic matter and salts have no effect on the color. Commercial potassium nitrite can be dissolved in alcohol, when, only the nitrite dissolves; the nitrate remains undissolved.

2. Acidify the water with a few drops of HCl before applying the tests. It takes a few minutes to develop the *red color*. (1) Dissolve aniline ($C_6H_5NH_2$) in Nordhausen acid, $H_2O2(SO_3)$, and you form sulphanilic acid, as follows: $C_6H_8NH_2 + H_2O2SO_3 = C_6H_7NSO_3 + H_2SO_4$. Use two drops. (2) Dissolve naphthaline ($C_{10}H_8$) in strong HNO_3, and you form nitronaphthaline ($C_{10}H_7NO_2$'. If this is treated with *ferrous acetate*, or with ammonia and hydrogen sulphide, it is reduced to naphthylamine, as follows: $C_{10}H_7NO_2 + 3H_2S = C_{10}H_9N + 2H_2O + S_3$. Use two drops. Add to the acidulated water, above described, two drops of sulphanilic acid, two drops of naphthylamine, when the water is colored red in the presence of nitrous acid.

the wind, and passes from earth to sky, and from sky to earth, and, anon, lies buried in its mould, in a short time to go on again in its ceaseless round to develop other forms of vegetation and life. Or he can view it in another light, and say that it has been the agent that has fitted the earth for the abode of man. It has been aptly compared to the blood — the grand motive power of the arts, the temperer of climate, the bounder of fertility, the locator of cities as well as trade and wealth, and the common carrier of the universe.

PART IV.

CHAPTER I.

EXAMINATION OF MILK.

Milk is a secretion of the mammary glands, and constitutes the entire food and drink of the young animal. It is usually confined to the mother, and also to the female sex; for exceptions, see Cazeaux on Midwifery, also Humbolt's Cosmos. For our present purpose, it may be described as a dilute solution of fat, milk-sugar, salts, caseine, and very small quantities of mineral matter. Chemists differ as to the reaction of fresh milk, for it may be acid or alkaline, usually alkaline. In the carnivora there is more fat and less sugar than in the herbivora. Ewe's milk is the richest, and cow's and human milk are about alike. Milk varies with (1) the time of the day, (2) kind of food, (3) the condition of the animal or person, (4) with the time since parturition, the colostrum being the richest, (6) in the same milking, the last portion being richer than the first — this has been observed in the human milk, (6) with the complexion of the person.

It has a pleasant and sweetish taste, and, when warm, an agreeable odor. The specific gravity of milk varies considerably, that of woman being sometimes as low as 1.020 (the average about 1.032), while that of the sheep is as high as 1.040. When the milk is allowed to stand, it separates into two portions — the upper layer of cream (10 to 14 per cent.), and a lower layer of mostly water.

COMPOSITION OF MILK (AFTER BOUSSINGAULT).

NAME.	HUMAN.	COW.	ASS.	GOAT.	MARE.	DOG.
Water	88.4	87.4	90.5	82.	89.63	66.30
Butter or fat	2.5	4.	1.4	4.5	Traces.	14.75
Sugar of milk and soluble salts	4.8	5.	6.4	4.5	8.75	2.95
Caseine albumen and insoluble salts	3.8	3.6	1.7	9.	1.60	16.
	99.5	100.	100.	100.	99.98	100.
Specific gravity	1.032	1.030	1.035	1.034	1.036	1.034

The milk, after a time, becomes acid; this is owing to the sugar ($C_{12}H_{24}O_{12}$) becoming converted into lactic acid ($H_2C_3H_4O_3$), as follows: $C_{12}H_{24}O_{12}) = 4(H_2C_3H_4O_3)$.

Condensed milk, when properly made, is a very good substitute for milk. It should be made by taking an ounce of sugar to a pint of milk, and boil it down in vacuum pans to about one-fifth of its original volume. Milk is coagulated by acids, tannin, alcohol, and wood spirit, also by alum and various other salts, and it is the most indigestible of all uncoagulated albuminoids. The best simple test for the adulteration of milk is to boil it; it should remain clear; adulterations thicken it, and sometimes settle at the bottom. Milk may be creamed in order to sell the cream separately; in this case it has more or less of a bluish color. To hide this color, flour, starch, chalk, caramel, and extract of chickory are added. The flour and starch can be detected by the microscope and by iodine; the chalk is found in the bottom of the vessel, and effervesces with hydrochloric acid: the coloring matters remain in the serum when the milk is coagulated, and can be detected. The macerated brains of sheep and other animals are said to be added to milk, but it would be hard for them to find all the brains they require for this purpose. There is such a difference between

normal milk, the richest and poorest, that a *little* water can
be added, or a *little cream* abstracted, and it cannot be detected
by chemical analysis. *Milk may be regarded as creamed, when the
proportion of butter is less than one-fifth of the total weight of solid
matter.*

ANALYSIS.

Water. — A weighed quantity of the milk is evaporated in a Pt
crucible on the water bath to dryness; when it has crusted over,
and it is nearly dry, it can be broken up, to expel the last traces
of water, with a Pt spatula. *The loss is water.* *The residue is the
milk solids,* which consists of ash, or mineral matters, fat, milk
sugar, and caseine.

Fat. — The milk solids of ten cubic centimeters of milk are
treated with ether, and heated to the boiling point; this is
repeated three or four times, pouring the ether through a small
filter if turbid, or into a large Pt dish, holding 100 cubic
centimeters. The last time, the residue may be moistened with
alcohol to soften it. The ethereal solution, about fifty cubic
centimeters, is evaporated in a warm place; as it approaches
dryness, it becomes turbid; it should be now transferred to a
water bath, and heated to 100° C. When the traces of water and
alcohol are expelled, it again becomes clear. The fat, or butter,
is now weighed; generally, this is all that is required in ordinary
milk analysis to determine milk solids and fat.

Caseine. — This is the nitrogenous constituent of the milk,
and it is thought by some to contain two distinct chemical
substances — caseine and albumen, the one (albumen) being
precipitated on boiling, the other, not coagulated. The caseine is
in two forms — soluble, in fresh milk, and insoluble, in sour
milk; this is in all probability due to molecular changes, like
soluble and insoluble silica. In milk the caseine is chemically
combined with phosphate of lime, and there is no known method
of separation, without destroying the caseine. The coagulation
of the milk is causing the caseine to become insoluble. The

residue, after the fat has been removed, is dried, and digested
with water, which dissolves the caseine and other soluble matters
of the milk. Alcohol precipitates the most of these soluble
matters (milk sugar), and leaves the caseine in solution, with
traces of milk sugar and soluble salts. It can be obtained purer
by treating skimmed milk with hydrochloric acid (HCl), collecting the curd upon a linen strainer, and washing it with water,
and *dilute* hydrochloric acid, and again with water. It is then
heated with a quite large volume of water at 110° F., and filtered,
and again reprecipitated by neutralization with ammonium
carbonate, $(NH_4)_2CO_3$. The precipitate is washed with water,
alcohol, and ether, to remove the last traces of fat; it is now as
pure as it can be obtained. As before stated, the ash, when
burned, always contains phosphate of lime.

Milk sugar $(C_{12}H_{22}O_{11})$.—The milk is coagulated by acid, the
caseine and fat removed as before described, the liquid, or whey,
evaporated to crystallization, and the crystals decolorized by
animal charcoal. Sugar of milk is sometimes called lactose, and
grape sugar is often called glucose. The sugar of milk can also
be approximately determined by Fehling's solution (see urine)
sixty-seven milligrams of milk sugar will reduce an amount
equal to fifty milligrams of grape sugar.

MILK SUGAR — $C_{12}H_{22}O_{11}H_2O$.	CANE SUGAR — $C_{12}H_{22}O_{11}$.
Specific gravity, 1.53.	Specific gravity, 1.60.
Reduces Fehling's solution.	Does not reduce Fehling's solution.
Dissolves six parts of water.	Dissolves in two parts of water.
If treated with strong H_2SO_4, no $\overline{SO_2}$.	If treated with strong H_2SO_4, - $\overline{SO_2}$.
Crystallizes in square prisms, and not very sweet.	Crystallizes in oblique rhombs, and very sweet.

The ash.—Evaporate the milk to dryness on the water bath;
heat the residue in a Pt crucible until the ash is *white*. It

consists mainly of phosphate of lime, which forms two-thirds of
it, and about one-third, chlorides. The analysis of milk by its
ash is important, as, by it, we can tell if the sample has been
watered. It, of course, requires a balance, but it can be done in
one hour, and we know of no *surer way* of determining if the
milk has been adulterated. During the months of January and
February, 1879, we made a number (twenty) of analyses of milk
from the Ohio State University farm, with reference to determining the amount of butter produced by different kinds of food
(a description of which would be out of place here, but can
be found in the Agricultural reports), and also its adulteration
with water. The analysis of the ash was as follows :

(1)	00.70 per cent.
(2)	00.70 per cent.
(3)	00.55 per cent.

You can see by the above analysis (3) which one was watered,
and also the per cent. of water added. It is well known that the
milk solids contain more ash than water. Milk exhibits great
constancy of composition, the effects of food showing itself in
the *amount* of secretion rather than in its *quality*. We are not
able to dilute the blood by administering water to the animal;
water tells on the perspiration and on the urine. Milk and
blood are alike in this respect, and differ from the urine. We
have not found the use of any of the instruments—lactometer,
etc., reliable in milk analysis.

Cheese consists mainly of caseine, salt, milk fat, some phosphates of lime, and water. It is made by coagulating the milk
by rennet (an infusion in water of the stomach of a young calf,
casually salted and dried to make it keep), then pressing the
curds, and afterward properly curing it. The analasis is as
follows :

For the *fat* and *caseine*, take ten grams (see that it is finely
divided) and boil it in a small flask with ether, and the solution
poured off; this is repeated three or four times, when the ethereal

solution is treated as for fat in milk. There remains behind, caseine, some milk sugar, salt, and phosphate of lime; this is now treated with strong alcohol, and then washed with boiling water, and dried. It consists of caseine and the small per cent. of phosphate of lime before spoken of (see milk analysis). It is now burned, and the loss is caseine; the ash is sometimes as high as 7 per cent., without being adulterated with mineral matter. If you suspect adulteration, examine the ash by the separation of the bases; sometimes oxides of the metals are added to make it weigh.

The milk sugar is determined by evaporating the alcoholic and aqueous solution, after the fat has been removed, to dryness, and igniting; the loss is the sugar, for it burns up.

The water is determined by taking about one gram, and drying it in a water bath until it ceases to lose weight; the ash can be determined by burning the residue.

PART V.

CHAPTER I.

EXAMINATION OF PTOMAINES.

The name "ptomaines" has been given by Selmi to bodies which have been detected in exhumed corpses, and resemble the vegetable alkaloids in their chemical reactions and physiloogical effects. These ptomaines are usually produeed by bodies, which after a brief exposure, have been excluded from the air, for they occur in buried bodies, sausages, and tinned foods, and, in most cases, the production is in the internal portion, though not in all cases. These ptomaines vary in their physiological actions; some act as poisons, some inactive, while others counteract the effect of poisonous substances.

Panum and Schwenigner have investigated the poisonous effects produced by food in certain stages of putrefaction or fermentation, and have found that different physiological actions are produced at different stages of decay; they have shown this to be true of albuminous substances. While Sonneschein and Zuelzer found in anatomical maceration fluid, an alkaloid which resembles atropine in its action; poisonous sausage produced similar effects; also, bodies which produced tetanic symptoms. In many cases of poisoning by food; as, for instance, cheese, the bad effects were not due to vegetable growths or microscopic organisms, and sometimes it is found quite fresh. Many individuals have derived from the putrefaction of maize, a body that will produce tetanic symptoms. The relation of these products of putrefaction to certain diseases is evident from the fact, that Sonneschein's alkaloid is found in the bodies of patients dying from typhus fever, and many persons poisoned by decaying food show marked typhus symptoms.

It becomes a matter of grave importance to the chemist to be able to distinguish between these poisonous bodies, which are the result of putrefactive processes, and the vegetable principles which, when administered, may produce death.

Brouardel and Boutmy propose the following test for distinguishing ptomaines from vegetable alkaloids :

The base, extracted from a dead body, is converted into a sulphate, and add a few drops of this solution to a small quantity of a solution of ferricyanide of potassium in a test tube, on adding *neutral* ferric chloride; if a ptomaine be present, Prussian blue will be formed; but if the base is a vegetable alkaloid, no reaction occurs.

But to this rule, there are many exceptions, and the worst thing about it is, many of the common vegetable poisons, notably, morphine and veratrine. To verify this statement, take a solution of the sulphates of morphine or veratrine, and add it to a *freshly* prepared solution of *pure* ferricyanide of potassium; now add by means of a glass rod, a part of a drop of neutral ferric chloride, when Prussian blue will be formed. Unless the poison is proven beyond the possibility of a doubt, it is best for the chemist to err on the side of mercy, and under *no circumstances* to allow professional reputation to be weighed in the balance with human life. Through the kindness of the officers of the Columbus Medical College, we made some ptomaines, and verified the statement above made in regard to morphine and veratrine.

It will be noticed that, while some alkaloids are confined to certain orders of plants, others again are of wide destribution, being found in many orders, and in nearly every part of the plant. To show the number of alkaloids, see the number obtained from the opium bases alone.

A FEW OF THE COMMON ALKALOIDS OF VEGETABLE ORIGIN.

NAME.	FORMULA.	ORDER.
Podophylline		Berberidaceae.
Emetine	$C_{30}H_{44}N_2O_8$	
Quinine	$C_{20}H_{24}N_2O_2$	Cinchonaceae.
Chinchonine	$C_{20}H_{24}N_2O$	
Aricine	$C_{23}H_{26}N_2O_4$	
Santonine	$C_{15}H_{18}O_3$	Compositae.
Strychnine	$C_{21}H_{22}N_2O_2$	Loganiaceae.
Brucine	$C_{23}H_{26}N_2O_4$	
Veratrine	$C_{32}H_{52}N_2O_8$	
Sabadilline } In small	$C_{41}H_{66}N_2O_{13}$	Melanthaceae.
Colchicine } quantities.	$C_{17}H_{19}NO_5$	
Jervine }		
Morphine	$C_{17}H_{19}NO_3$	
Codeine	$C_{18}H_{21}NO_3$	
Thebaine	$C_{19}H_{21}NO_3$	
Narcotine	$C_{22}H_{23}NO_7$	Papaveraceae.
Papaverine	$C_{20}H_{21}NO_4$	
Narceine	$C_{23}H_{29}NO_9$	
Aconitine	$C_{30}H_{47}NO_7$	
[One of the rankest poisons known; fatal dose 1-10 grain.]		Ranunculaceae.
Macrotine		
Delphinine	$C_{24}H_{35}NO_2$	
Atropine	$C_{17}H_{23}NO_3$	
Daturine	$C_{17}H_{23}NO_3$	Solanaceae.
Hyosciamine	$C_{17}H_{23}NO_3$	
Nicotine	$C_{10}H_{14}N_2$	
Solanine	$C_{42}H_{75}NO_{15}$ (GLUCOSIDES.)	
Conine	$C_8H_{15}N$	Umbelleferae.
Caffeine	$C_8H_{10}N_4O_2$	Rubiaceae.
Theine	$C_8H_{10}N_4O_2$	Ternstroemaceae.
Theobromine	$C_7H_8N_4O_2$	Rubeaceae.
Leptandrine		Scrophalariaceae.
Pilocarpine	$C_{11}H_{16}N_2O_2$	Rutaceae.

ANIMAL ALKALOIDS.

NAME.	SYMBOL.	SOURCE.
Xanthine	$C_5H_4N_4O_2$	Urine.
Hypozanthine	$C_5H_4N_4O$	Flesh.
Kreatine	$C_4H_7N_3O$	Flesh, Urine, Brain.
Sarcosine	$C_3H_7NO_2$	Flesh, Urine, Brain.
Urea	CH_4N_2O	Urine.

OPIUM BASES.

Morphine	$C_{17}H_{19}NO_3$	
Codeine	$C_{18}H_{21}NO_3$	
Codamine	$C_{20}H_{23}NO_4$	
Laudanine	$C_{20}H_{25}NO_4$	
Pseudomorphine	$C_{17}H_{19}NO_4$	
Thebaine	$C_{19}H_{21}NO_3$	} Isomeric.
Thebenine	$C_{19}H_{21}NO_3$	
Protopine	$C_{20}H_{19}NO_5$	
Papaverine	$C_{21}H_{21}NO_4$	
Deuteropine	$C_{20}H_{21}NO_5$	
Cryptopine	$C_{21}H_{23}NO_5$	
Meconidine	$C_{21}H_{23}NO_4$	
Laudanosine	$C_{21}H_{27}NO_4$	
Rhoeadine	$C_{21}H_{21}NO_6$	} Isometric.
Rhoeagenine	$C_{21}H_{21}NO_6$	
Narcotine	$C_{22}H_{23}NO_7$	
Narceine	$C_{23}H_{29}NO_9$	
Lanthopine	$C_{23}H_{25}NO_4$	

CHAPTER   II.

EXAMINATION   OF   POISONS.

A poison is any animal, vegetable, or mineral substance which, when applied externally, or taken into the stomach or circulatory system, operates such a change in the animal economy as to produce disease or death. Its action may be chemical or physiological. * The physiological may be — (1) Corrosive, (2) Irritant, (3) Neurotic. The chemical, (1) Organic, (2) Inorganic.

The action of the poisons may be local or remote; local when confined to the part where the poison is applied, as in the case of strong acids, etc.; remote when the action extends to distant organs. Poisons have a wonderful power of selecting organs or tissues peculiar to themselves — strychnine, the spinal cord; opium, the brain; hydrocyanic acid, the lungs; digitalis, the heart. Death is, in most cases, due to remote action. The action of the poison on the system is modified by — (1) the quantity, (2) the molecular or physical condition of the poison — most active as a gas or vapor, next as a liquid, and least as a solid — (3) chemical combination. (*a*) It may be increased when morphine is given with acetic or hydrochloric acid. (*b*) It may be diminished when sulphuric acid and caustic soda are given together. (4) Mechanical mixture, when its action may be

---

* Dr. Bennett has shown in the British Medical Journal that too much reliance cannot be placed upon physiological antidotes, and especially upon experiments on animals. The power of a hog to resist snake bites is proverbial. A dog will take a large quantity of arsenic, or a rabbit, belladonna; but a small quantity of belladonna given to a dog, or of arsenic to a rabbit, will produce death. He has shown that dogs will take atropine, and horses strychnine in *very large doses*. When poison is used to destroy vermin, it should never be placed upon the ground. The power of fresh earth to absorb poison is very great.

delayed. (5) The manner of giving it by the mouth, wound, etc. (6) The habits of the person; and closely connected with this is the idiosyncrasy of the person. (7) The condition of the person as to health. It is retarded by sleep and by food.

The evidences of poisoning are—(1) the symptoms, (2) post mortem appearances, and (3) chemical analysis. As the symptoms of many poisons and diseases are alike, the burden of proof remains for chemical analysis. Before we take up the chemical analysis of poisons a few words might be said as to general treatment: (1) Get the poison out of the system as soon as you can; (2) If you cannot remove it, neutralize it if possible; (3) Assist its elimination by cathartics and diuretics; (4) Treat the symptoms, as they arise, upon general principles. The first may be secured by emetics and the stomach pump. The second requires antidotes, of which there are three classes —chemical, mechanical, and physiological. The following are considered as the requisites of an antidote: (1) It should be easily accessible; (2) It should be capable of being administered by any one. It should be capable of being taken in *very large quantities without injury*. The third and fourth require no special explanation, but the ordinary good judgment of the practitioner. The organs examined are usually the stomach and liver. (1) It is recomended that the student take the pure poison (say strychnine), and see that he can confirm it by all the tests given in the book; (2) Add it to food, making it as near as possible like the contents of the stomach, and then try to *separate* and *indentify* it; (3) To take some small animal, as a cat or dog, and administer a poisonous dose, and to notice the following: (1) The quantity you take, which is always weighed; (2) The length of time before symptoms of poisoning begin; (3) The duration of the symptoms until recovery or death; (4) The character of the symptoms; (5) Post mortem appearances. If, for instance, you were told the person had violent convulsions; that his mind remained clear; that, at the post mortem, the hands were clinched, the whole body was rigid, and opisthotonos resting, so to speak, on the head and heals, it would be reasonable to *infer* and *examine for strychnine ; but this should not exclude the examination for other poisons.*

The first requisite of analysis is cleanliness. See that everything is *absolutely clean* by washing it first with acid, and rinsing it with water three or four times; then again washing it with an alkali (as NaHO or KHO), and again rinsing it with water three or four times, and lastly, with distilled water a couple of times; *also, test the purity of every reagent used.* In most every step of your work you will find many difficulties, which it is impossible to describe; and the remedies for which must be suggested by your general knowledge of chemistry.

The inorganic poisons are generally quite easily detected, and will not usually give any trouble. The organic poisons are more difficult and more readily pass off, in some cases leaving no traces behind. They are first examined for this reason. In all cases only one-third to one-half of the contents of the stomach, or other organ, is taken; the rest is kept in case of any accident in the work. Take one-half of the contents of the stomach, or other organ, strain it through a linnen cloth, and examine the solid materials with a good hand lens. If nothing is found, mix the contents well together, put in a new or *clean* wide-mouth bottle, and keep at a warm (75° C.) temperature, and examine for prussic acid by the escaping vapors coming in contact with the reagent in the mouth of the bottle, the bottle being covered by a clean glass plate. For the tests, see prussic acid ; the odor is sometimes characteristic; a few hours are usually sufficient for this test. The contents to be examined are now transferred to a retort, mixed with a small quantity of distilled water, and one-fifth distilled over into a receiver by means of the heat of a sand or water bath; this can be examined for hydrocyanic or prussic acid, and the distillation continued almost to dryness, when volatile poisons — the ones we are especially interested in, can often be detected by their odor — ether, turpentine, alcohol, etc., and must be examined by their appropriate tests. If there is no evidence of any volatile poison, the distillate is returned to the retort, and a few drops of *pure* sulphuric acid is added, and the operation repeated, this time placing a little silver nitrate, caustic potash, or caustic soda in the receiver, to catch any acid that may come over. If this is done in the dark, phosphorus, if present, may be identified. The potash solution is again tested for hydrocyanic acid or other poisons. The contents of the retort is now treated with three times its bulk of strong alcohol, and placed in a warm place for twelve hours, occasionally shaking it up. Filter, and separate the filtrate into two portions —(a) and (b), but first carefully preserve the solid matter on the filter paper * for future reference, and make this an invariable rule, never throw anything away, but always put it away carefully, labelled, and where in the analysis it was obtained.

Test (a) for the mineral poisons by the separation of the bases before given. If arsenic or other mineral poisons are found, it would be well to commence over again, taking one-third of the stomach and its contents, cut up into small pieces, treat with hydrochloric acid, either in a retort or in a glass flask (because the chlorides of the metals are volatile), a number of times; by this means you obtain all the metals as chlorides, and the estimation of the quantity can be easily found. But if nothing is found in (a) by sulphuretted hydrogen, remembering the precautions given in the separation of the bases, the solution is boiled to expel the sulphuretted hydrogen, and it is mixed with (b), and an excess of lead acetate is added, and filtered. (1) The organic and other acids combine with the lead, forming their respective lead salts. The precipitate contains the lead in combination with acids; this is added to water, and H_2S passed through, when the lead is precipitated as a sulphide (PbS), and the acids examined in the solution. The filtrate

---

* This is cut up very fine, and pulverized in a mortar, treated with three parts of water and one of hydrochloric acid, and brought to about 90 degrees for one-half hour, the solution filtered off, and it again treated in the same way; the solutions are now concentrated on the water bath, and pure copper foil is added. If there is any deposit, it is examined for mercury, arsenic, or antimony, by their respective tests. To be sure that everything is dissolved with the hydrochloric acid, the residue is incinerated in a porcelain crucible, and treated with nitric acid; dilute, and treat with sulphuretted hydrogen. In some cases, lead is found.

(1) contains the excess of lead acetate; H_2S is passed through to precipitate the lead as PbS; now it is filtered, the filtrate is evaporated on the water bath almost to dryness, and treated with water and a *few* drops of acetic acid; to this add a few drachms of water, and filter, again evaporate almost to dryness, and add a solution of pure caustic soda; · it is now treated with ether, and well shaken. The solution of ether is removed by a pipette (the pipette can be extemporized by taking a soft piece of rubber tubing three inches long, plugging up one end with a glass rod, and fitting the other end to a small glass tube); this operation may be repeated three or four times. Chloroform can take the place of the ether in some cases, as it is a better solvent for morphine and cinchonine. It is well here to know the taste of the most important alkaloids, and to taste the solution. You can now evaporate or distil the ether or chloroform, and examine the residue for alkaloids by adding a few drops of acetic acid; the filtering of it is necessary to free it from fatty matters. Except in cases of actual poisoning when human life is at stake, this analysis may be very much shortened: (1) Test for volatile poisons by distillation; (2) treat the finely divided contents of the stomach or other organs or tissues with acetic acid, enough to make it quite strongly acid, and water enough to make it a pasty mass; now treat it with one-half its volume of strong alcohol, and digest on the water bath, with frequent stirring for one hour; strain through a linen cloth, and well wash the residue with strong alcohol; concentrate the solution, and treat with ether or chloroform, or both, and examine for the alkaloids.* The residue is treated with hydrochloric acid a number

---

*The alkaloids are divided into two divisions — volatile and non-volatile. To the first belong *nicotine* and *conine*. The second has three subdivisions: first, those that are precipitated by potash or soda or a solution of their salts, and redissolves in an excess of the precipitant — morphine is the most important one; second, those precipitated by potash or soda, but do not redissolve to any great extent by an excess of the precipitant, and are precipitated by bicarbonate of soda from acid solutions — *narcotine*, *quinine*, and *chinchonine*; third, those precipitated by potash or soda, and do not redissolve to any great extent by an excess of the precipitant, but are not precipitated by a bicarbonate of the fixed alkaline metals — *strychnine*, *brucine*, *veratrine*, and *atropine*.

of times as described in the foot note on page 123. The following is the method of Selmi for the detection of alkaloids:

The organic substance is digested for three or four hours with alcohol and sulphuric acid, filtered, and the residue again and again treated in the same way, the filtrates mixed and evaporated to a syrup, and treated with freshly prepared barium hydrate. After the addition of anhydrous baryta and powdered glass, the whole is reduced in a mortar to a coarse powder, and shaken up with ether, and the filtrate digested with freshly prepared lead hydrate; by now treating with ether, the alkaloid is obtained quite pure.

Strychnine ($C_{21}H_{22}N_2O_2$) was discovered in 1818 by Pelletier and Caventon. It is found in the *strychnos nux vomica* and the St. Ignatius' bean — seed of the strychnos Ignatia. The quantity varies from .5 to 1 per cent. Brucine is often used to adulterate, and occurs with it in the bean. The presence of brucine may be known by nitric acid giving a blood red color, while it gives no coloration with strychnine. The best solvent is chloroform; the medicinal dose is from one-thirtieth to one-twelfth of a grain, the smallest fatal dose is from one-fourth to one-half of a grain. When taken in poisonous doses, the symptoms come on suddenly; the patient has violent tetanic convulsions; the pain, intense; the pulse, rapid; the mind, clear; and vomiting is not common. The post mortem appearances are engorgement of the lungs, congestion of the brain and spinal cord, and body opisthotonos. The treatment is by emetics and the stomach pump, when they can be used; the use of chloroform, tannic acid, opium, camphor, and chloral hydrate have been recommended.

TESTS.

(1) It is white, and with an intensely bitter taste; when heated on Pt foil, it melts and burns like resin, with a black smoky flame.

(2) It is easily dissolved in dilute acids, forming very soluble salts, and is precipitated from its solution by fixed caustic alkalies, and ammonia.

(*3*) It dissolves in nitric acid with a red coloration, due, possibly, to brucine; if dissolved with sulphuric acid, it is colorless. If to this solution you add oxygen or any oxidizing agent; as, a small crystal of bichromate of potash, ferricyanide of potassium, black oxide of manganese, or peroxide of lead, you will get a beautiful play of colors—purple, violet, and crimson. Black oxide of manganese is the best reagent for the above, and potassium bichromate is the worst. Dr. Letheby uses a galvanic battery to generate the oxygen; curarine gives the same reaction with sulphuric acid and potassium chromate, but curarine is colored by sulphuric acid alone, while strychnine is not.* Morphine interferes with this reaction. Sulphomolybdic and iodic acids produce no immediate change of color in strychnine, while they do in morphine (see morphine).

(*4*) Sodium bicarbonate, potassium sulphocyanate, mercuric chloride, or iodide of potassium produces a white precipitate.

(*5*) Bichromate of potash produces a bright yellow precipitate. The chlorides of platinum, also the chlorides of gold, produce yellow precipitates. The aqueous solution of iodine in potassium iodide produces a reddish brown precipitate.

(*6*) Dr. Hall's physiological test is made by injecting a dilute solution of strychnine into the thorax or abdominal cavity of a frog, causing tetanic convulsions. Many of the tests are delicate, when examined with a microscope.

---

*Curarine gives with sulphuric, a red color; strychnine gives none. Curarine may be separated from strychnine by means of benzene, in which the former is insoluble. Cod liver oil gives a similar reaction, distinguished by the taste. Aniline, pyrozanthine, papaverine, narceine, veratrine, and solanine belong to this list, also.

Brucine ($C_{23}H_{26}N_2O_4$) was discovered by Pelletier and Caventou in 1819. It is usually found with strychnine and nux vomica. It takes from six to ten times as much brucine as strychnine to produce death. The symptoms and mode of treatment are alike, those of brucine coming on more slowly and less violent than strychnine. Chloroform is one of the best solvents. In separating it from organic mixtures, the same process is used as for strychnine (see the general method for alkaloids. page 122).

TESTS.

(*1*) The residue is tested with nitric acid; a red color indicates brucine; if it is heated, it changes to yellow; now add a trace of protochloride of tin, when the color changes to a deep purple; an excess of acid or tin bleaches it.

(*2*) When mixed with sulphuric acid and bichromate of potash, orange, green, and yellow tints are produced in turn, due to the reduction of the chromium, and differing from strychnine.

(*3*) A solution of iodine in iodide of potash gives with brucine an orange brown amorphous precipitate, insoluble in acetic acid.

(*4*) Bichloride of platinum gives a yellow precipitate of the double chloride of platinum and brucine, soluble in caustic alkalies, and insoluble in acetic acid. The chloride of gold gives a somewhat similar amorphous precipitate.

Morphine ($C_{17}H_{19}NO_3$) was discovered by Serturner in 1804. It is the poisonous alkaloid of opium, good opium yielding from 6 to 8 per cent. of the alkaloid. There are other alkaloids found in opium, the most important of which are narcotine and codeine. The best solvent is acetic ether. The medicinal dose is one-fifth of a grain; the smallest fatal dose is one grain. It is about six times as strong as good opium; drowsiness and stupor are the first symptoms; pulse, weak; breathing, slow and almost imperceptible. The pupils, in most cases, are contracted; when large doses (poisonous) are given, it is sometimes attended with convulsions. The post mortem appearances have nothing characteristic, except, in some cases, the peculiar odor of opium. During treatment, keep the patient constantly aroused by any means. The poison should be eliminated by emetics and the stomach pump. If the poison can be removed, in some cases the magneto-electricity is used with good effect to prevent insensibility; solutions of iodine, bromine, and tannic acid have been recommended; strong tea or coffee is the best drink. Ammonia can be applied to the nostrils in case of a collapse. The existence of opium is determined by the presence of morphine and meconic acid.

TESTS.

(*1*) Nitric acid gives, at first, a bright red, then an orange red color; chloride of tin does not decrease the color, differing from brucine.

(*2*) Iodic acid gives a brown color, a very delicate test. If iodic acid and bisulphide of carbon are mixed together, no change of color occurs; if we add morphine (solid or solution), iodine is separated from the iodic acid, and dissolves in the bisulphide, coloring it pink or red; of course, a solution of starch will do as well, giving a blue color.

(*3*) Dissolve ammonium molybdate in strong alcohol, and it should be made as required for use. It gives with morphine or its salts a reddish purple or crimson red color; this changes to a green, and, ultimately, to a sapphire blue.

(*4*) Sulphuric acid gives with morphine no change of color, giving a yellow color with narcotine; if bichromate of potash is added, a bright green color is given from the reduction of the chromium salt.

(5) Meconic acid ($C_7H_4O_7$) gives a red color with the perchloride of iron. Sulphocyanides, acetic acid, and neutral acetates give the same reaction with iron, but the sulphocyanides are discharged by corrosive sublimate; acetic acid and neutral acetates give no precipitate with acetate of lead, while meconic acid does. Chloride of barium gives a white crystalline precipitate; nitrate of silver, a yellow amorphous precipitate.

Under this head might be included laudanum, opium, paregoric, soothing syrup (Mrs. Winslow's), pulmonic wafers (Locock's), cordials (Godfrey's), etc. Concentrated sulphuric acid, to which a *small* per cent. of nitric acid has been added, gives a violet purple color, when gently heated with morphine or its hydrochlorate.

Codeine ($C_{18}H_{21}NO_3$) was discovered by Robiquet in 1832; opium contains about 1 per cent. It resembles, in a general way, morphine, but can be distinguished by the tests.

TESTS.

(1) It differs from morphine in not decomposing iodic acid.

(2) It does not give any red color with nitric acid; it differs from narcotine in not turning yellow, but of a light pinkish brown color by sulphuric acid; it resembles both morphine and narcotine in its reactions with chromium.

Narcotine ($C_{22}H_{24}NO_7$) was discovered by Derosne in 1803; opium contains from 6 to 8 per cent. It is thought by many that these three alkaloids are combined more or less together, their physiological effects being about the same. It is not commonly sought for in medico-legal investigations. It is dissolved by boiling alcohol; the ether solutions are very bitter. Nitric acid gives a *yellow*, and not an orange red like morphine; sulphuric acid gives a sulphur yellow color, but with morphine, a pinkish brown; they agree in the chromium test. If to sulphuric acid and narcotine, a grain of saltpetre is added, a deep blood red color is obtained, but not with morphine. It does not decompose iodic acid, while morphine does.

Quinine ($C_{20}H_{24}N_2O_2$) was discovered by the Spanish conquerors of Peru in the early part of the seventeenth century, but the alkaloid was perfected by Pelletier and Caventon, in 1820. Its physiological action is as follows: the pulse rate falls; the coagulability of the blood is lessened; it deranges, enfeebles, and finally extinguishes nervous action; the animal staggers, becomes agitated and sometimes convulsed, and then assumes a dull, inanimate expression; the vision is impaired, and the pupils are widely dilated. When given in large doses, the general sensibility is obtuse, and the limbs tremulous; also, the patient is accompanied by great depressive apathy, somnolescence, unsteadiness of gait, and impaired sight and hearing. As death approaches, the skin loses its sensibility, and the limbs their power of motion. Like many other medicines, quinine is stimulant in small, and sedative in large doses; but it differs from the other stimulants in the duration of its action, which is long sustained, and entitles it to be called a tonic stimulant; the dose is from two to eight grains. *It is questioned if quinine alone has ever produced death.*

TESTS.

(*1*) Most alkalies precipitate it as a hydrate, which may be crystalline.

(*2*) An aqueous solution of the acid salts exhibit a blue color by reflected light; the solution is intensely bitter.

(*3*) Add chlorine water to the solution of a salt of quinine; if ammonia is now added, it gives an emerald green color; quinodine gives the same reaction. If after the addition of the chlorine water, a solution of ferrocyanide of potassium is added, and now a few drops of ammonia or most any other alkali, when you will get a deep red color, changing to a dirty brown; the red color will vanish, if you add a drop of acetic acid.

(4) HESSE'S TEST.—Take .5 gram of the sulphate; dissolve in ten cubic centimeters of water, warmed to 60° C. (140° F.). After ten minutes, the cooled liquid is filtered, and five cubic centimeters of it are slowly agitated with one cubic centimeter of ether.; after the separation of the ether, both strata should be clear. Cinchonine and cinchonidine will remain undissolved.

(5) Take *cinchonine* salts, must be nearly neutral, and add a solution of potassium ferrocyanide, when a flocculent precipitate of cinchonine ferrocyanide is formed; now add a slight excess of the ferrocyanide, and heat *slowly* and *gently*, when the precipitate dissolves, but separates again, upon cooling, in golden yellow scales, and can be best seen under the microscope; this is a very delicate, and characteristic test.

Veratrine ($C_{32}H_{52}N_2O_8$) was discovered by Meissner in 1819. About thirty grains are obtained from one pound of the seed of sabadilla, or cevadilla. The ordinary medicinal dose of the commercial alkaloid is about one-sixteenth of a grain. It has no odor, but when applied to the nostrils, it produces violent sneezing; when an alcoholic solution is applied to the skin, it produces a prickling sensation. Alcohol, benzol, dilute acids, and chloroform are good solvents. It is not much used for criminal purposes. The symptoms are vomiting, convulsions, and insensibility. Treatment: remove the poison from the stomach as soon as possible; tannic acid and opium are recommended with good results in some cases. There is no known chemical antidote. Selmi has found ptomaines in the viscera that act like this poison; see ptomaines, page 117.

(*1*) Strong sulphuric acid gives a yellow color, changing to a reddish tint, and finally becomes a ·crimson red color; heat accelerates this reaction. Hydrochloric acid gives no color, but, if heated, it becomes red, resembling the permanganate of potash.

(*2*) Bromine in hydrobromic acid gives a dirty yellow color, a very delicate test.

(*3*) Iodine in iodide of potassium gives a reddish yellow color, also, a very delicate test.

(*4*) Picric acid gives a greenish yellow precipitate.

(*5*) Platinum chloride and ferricyanide of potassium give dirty yellow precipitates; ferrocyanide gives no precipitate.

(*6*) Tannic acid gives a white flocculent precipitate in quite dilute solutions.

Aconitine ($C_{30}H_{47}NO_7$) is, possibly, the most powerful poison known. It was discovered by Geiger and Hesse in 1832. A pound of the dried root (aconitum napellus) will yield from twelve to thirty-six grains, or from .1 to .2 per cent. Alcohol, chloroform, and benzol are its best solvents. The medicinal dose is one-130th part of a grain, and is rarely given in the form of the alkaloid; it is doubtful if the alkaloid can be administered internally with safety; one-tenth of a grain can be regarded as a fatal dose. When taken in poisonous doses, the symptoms come on rapidly: there is diminished sensibility; the skin loses its *sensation*, whilst there is deafness, and ringing in the ears, dimness and loss of sight; the pulse is low, feeble, and irregular, becoming at last almost imperceptible, with clammy cold sweats; at last, after a few convulsive gasps, the patient expires. In treating, evacuate the contents of the stomach by the stomach pump and emetics; such as, sulphate of zinc; brandy and ammonia may be used as stimulants. The following substances are recommended — vegetable infusions containing tannic acid, iodine in iodide of potassium, also, a dilute solution of nux vomica; strong tea and coffee can be given. The post mortem appearances are characterized by general venous congestion; the brain, liver, and lungs being more or less engorged, and usually accompanied by signs of gastro-intestinal irritation.

TESTS.

(*1*) The physiological action is its chief test; if you rub it on the inside of the gums, it produces a sense of tickling and numbness.

(*2*) When administered with a hypodermic syringe on small animals, it produces symptoms as above described.

(*3*) Chloride of gold gives a yellow precipitate.

(*4*) Iodine in potassium iodide gives a reddish brown precipitate.

(*5*) Picric acid (carbazotic) gives a yellow precipitate.

NOTE.—The small quantity required to produce death, symptoms, and physiological action, are of as much importance as the chemical tests.

Atropine ($C_{17}H_{23}NO_3$) was first announced by Brandes in 1819, and in 1833 by Mein, a German pharmaceutist, who obtained it pure. It is found in roots, leaves, and berries of the *atropa belladonna*, or *deadly nightshade*, in the proportion of a grain to the ounce of the roots, or less than .5 per cent. The fatal dose is two grains; the medicinal dose for hypodermic injection should not exceed one-250th of a grain. It is soluble in alcohol, ether, chloroform, or benzol. Symptoms—the patient is drowsy and giddy; the pulse is strong and rapid, the action of the heart being increased; the eyes are prominent and sparkling, and the pupils always dilated. There is often a desire to micturate or walk, and an inability to do either. As it approaches fatal termination, there is delirium and sometimes convulsions; these may alternate, and either may end in death. The post mortem appearances are not well marked. The treatment is by emetics and the stomach pump; tannic acid, iodine in potassium iodide, or one-fifth of a grain of morphine, administered hypodermically, may be used to keep the patient at rest; if the patient has taken an emetic, and it has operated, a good dose of castor oil and strong coffee may be given without harm.

It, of course, will be understood that belladonna acts like atropine, and requires the same treatment.

TESTS.

(*1*) When treated with the chloride of potassium, and mercury,* it gives a dense, white precipitate in very dilute solutions.

(*2*) Bromine in hydrobromic acid gives a yellow precipitate.

(*3*) Iodine in iodide of potassium gives a yellow precipitate.

(*4*) Chloride of gold gives a citron yellow precipitate.

(*5*) Tannic acid gives a white amorphous precipitate.

(*6*) Physiological tests are of great importance—dilating the pupil, but it must be borne in mind that daturine (stramonium) and hyoscyamine (Henbane) produce a similar result, only in a lesser degree.

Nicotine $(C_{10}H_{14}N_2)$ is found in the common tobacco plant, of which it contains from 4 to 8 per cent. It was discovered by Posselt and Reimann in 1828. It is a transparent, colorless, oily liquid, and one of the most active poisons known. It is soluble in water, alcohol, ether, chloroform, turpentine, and fixed oils;

---

*Made by dissolving 13.55 grams of corrosive sublimate and five grams of iodide of potassium, in one liter of distilled water The precipitate consists of an insoluble compound of an hydriodate of the alkaloid with an iodide of mercury. It precipitates albuminous substances, which should be first removed; the precipitates are insoluble in acids (distinctive from ammonia) or in dilute alkalies. It is one of the most delicate tests for the presence of any alkaloid. The organic liquid should be filtered and, if much colored, dialyzed. It may form an explosive compound with ammonia and an organic liquid. This reagent can be made a quantitative test for many of the alkaloids, the precipitate being mostly yellowish white; it is known as Mayer's reagent, and gives with nearly all alkaloids, except caffeine, colchicine, digitaline, and theobromine, the above described precipitate.

chloroform and ether extract it from its aqueous solutions. The characteristic symptoms are vertigo, nausea, vomiting, extreme prostration, trembling of the limbs, etc.; respiration is difficult, the skin cold and clammy. The poison is very rapid in its action, and when taken in poisonous doses (one drop usually fatal), death occurs in a few minutes, even equalling hydrocyanic acid in the rapidity of its action. In treating, remove the poison by emetics, etc. Afterwards allay the pain with opium or its equivalent, and preserve power with stimulants. The post mortem appearances are not very characteristic. The poison should be looked for in the stomach, liver, and lungs.*

TESTS.

(1) It gives many tests that are identical with ammonia; as, chloride of platinum, corrosive sublimate, arsenic, and nitrate of silver, but can be *distinguished from ammonia by its odor.*

(2) Picric acid gives a yellow precipitate.

(3) Chloride of gold gives a yellow precipitate, insoluble in acetic and hydrochloric acid, but soluble in caustic alkalies.

(4) Iodine in iodide of potassium gives a reddish brown precipitate, soluble in alcohol and in potash, while this reagent gives no precipitate with ammonia.

(5) The chlorides of potassium and mercury (see page 134) give a copious precipitate, even in dilute solutions. The action of a solution of the residue on small animals should never be omitted.

Conine ($C_8H_{15}N$) is an alkaloid from the common hemlock; the death of Socrates is generally believed to have been due

---

* In some recent experiments on nicotine by Kissling, the author examined tobacco smoke, and found carbonic oxide, sulphuretted hydrogen, hydrocyanic acid, picoline bases and nicotine are the most active poisons. The first three are in small quantities, and their volatility is too great to be of importance. The picoline are in very small quantities. The toxic action is due to nicotine; the proportion destroyed by the combustion of the cigar is quite small.

to this poison. The alkaloid is most abundant in the fruit of the plant, containing about 1 per cent. It was first obtained as an impure sulphate by Giseke, in 1827. One drop may be regarded as a fatal dose. The symptoms are a gradual and complete paralysis of the extremities, enlargement of the pupils, and loss of power. The paralysis gradually extends to the muscles of respiration, and the patient dies by apnœa. Nicotine and conine are volatile alkaloids, and are liquid at ordinary temperatures; they are colorless, oily, and volatile.

NICOTINE.	CONINE.
Tobacco odor.	"Mousy" odor.
Freely soluble in water.	Sparingly soluble in water.
No crystals with HCl fumes.	HCl fumes give crystals.
AgNO₃ gives a white precipitate.	AgNO₃ gives a dark brown precipitate.

The post mortem appearances are characterized by the *stomach* being congested, the *lungs* invariably so, the *intestines* healthy, the *brain* more or less congested, and the *blood* fluid. In treating, remove the poison by emetics and the stomach pump; then give stimulants.

TESTS.

(*1*) Corrosive sublimate gives a white amorphous precipitate.

(*2*) Nitrate of silver gives a dark brown and afterward changes to the black suboxide.

(*3*) Tannic acid gives a dirty white precipitate, soluble in hydrochloric acid.

(*4*) Iodine in iodide of potassium gives reddish amorphous precipitate, a very delicate test.

(*5*) Picric acid gives a yellow precipitate.

(*6*) Chloride of gold gives a yellowish white precipitate, but platinum chloride gives no precipitate.

Caffeine, or theine, ($C_8H_{10}N_4O_2$) is found in coffee, tea, or mate. Guarana, the dried paste of the fruit of the *Paullinia Sorbilis*, contains 5 per cent. This alkaloid was first separated in an impure state by Pilletier and Caventon, Robiquet and Runge, in 1821. Oudry found it in tea in 1827. Mulder Jobst, in 1838, showed caffeine and theine were identical. There is no record of death in the human subject. It is soluble in 100 parts of cold water; freely soluble in hot water, and in water acidulated with an acid; slightly soluble in cold alcohol, and tastes slightly bitter, and possesses feeble basic properties. With sulphuric and hydrochloric acids, it forms crystallizable compounds. In its physiological action, it excites the heart and respiratory movements, increases arterial tension, and arrests the rapid consumption of tissues. In the language of Cowper, it is

"The cup that cheers, but not inebriates."

Concentrated nitric and, also, sulphuric acid dissolve it without change of color.

TESTS.

(*1*) Moisten the residue, supposed to contain caffeine, with hydrochloric acid, and add a small crystal of chlorate of potash; now heat on the water bath for a few minutes, after which expose to the fumes of ammonia, avoiding an excess; a crimson or purple color proves the presence of caffeine.* Or it can be moistened

---

*The student is reminded of the similarity of this test and the one for uric acid on page 85. The following formula will explain the reason:

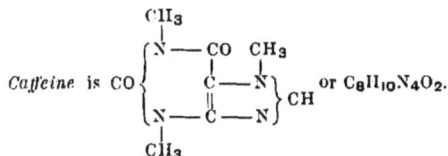

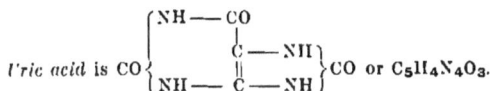

$$CH_3$$

Caffeine is $CO\begin{cases} N\text{---}CO\ CH_3 \\ C\text{---}N \\ N\text{---}C\text{---}N \end{cases}CH$ or $C_8H_{10}N_4O_2$.

$$CH_3$$

Uric acid is $CO\begin{cases} NH\text{---}CO \\ C\text{---}NH \\ NH\text{---}C\text{---}NH \end{cases}CO$ or $C_5H_4N_4O_3$.

What a singular example of the so called "brick dust" deposit of the urine, resembling the stimulating principle of the tea and coffee.

with nitric acid, evaporated to dryness on the water
bath, and treated with ammonia.

(2) Gold chloride gives a yellow precipitate.

THE FOLLOWING ARE VERY DELICATE TESTS FOR THE ALKALOIDS.

Mayer's Test.— See page 134.

Potassiobismuthous Iodide.— Precipitate the bismuth by sulphuretted hydrogen; dry, and treat with iodine in a large flask
at a gentle heat, when the iodine takes the place of the sulphur;
now take sixteen parts of liquid bismuth iodide, three parts
of potassium iodide, and three parts of hydrochloric acid.
*Always test your reagent to see that it is not decomposed by water
alone.* An orange colored precipitate is given by most alkaloids.

Marme's Potassio Cadmic Iodide.— Take twenty parts or
grams of iodide of cadmium,* forty parts or grams of iodide of
potassium, in 120 cubic centimeters of water, gives a precipitate
varying from gray yellow to yellow.†

Scheibler's Metatungstic Acid.— Add phosphoric acid to a
solution of ordinary tungstate of sodium as long as a precipitate
is formed and redissolved. The precipitates are white and
flocculent. You can detect one part of strychnine in 200,000
parts of water with this test.

Iodine in Iodide of Potassium.— Take three grains (one-fifth
gram) of iodide of potassium, and dissolve in one drachm (four
cubic centimeters) of distilled water; now add one grain (one-
fifteenth gram) of pure iodine. This is a very delicate test.

A very nice method of separating the alkaloids is due to

---

*Cadmium iodide is made by digesting a piece of cadmium with iodine in water
at a gentle heat for a number of hours; filter, and evaporate to dryness on the water
bath.

†The solution of the alkaloid should be *feebly* acidulated with sulphuric acid
before applying the reagent.

Prof. Prescott by means of treatment by water, ether, and chloroform. See Organic Analysis, page 138.

The common acids can easily be detected by the tests given in separation of the bases (As and Cr as acid) and the separation of the acids. They are common articles of trade, and can be procured without suspicion; most cases of poisoning are usually accidental or suicidal. The mouth is the part commonly affected, but if the spoon has been put far back, the mouth may escape. Death may take place from asphyxia. The pain is intense; the thirst, great; the patient cannot swallow, speak, nor scarcely breathe. Death may in many cases be due to starvation, as the patient cannot take food. In treating, neutralize the poison by giving the white of egg, soap and water, chalk and water, or calcined magnesia. If nothing handy can be obtained, plastering from the ceiling or wall may be used. Nutritive enemata must be given, if other means of administering food fails. The post mortem appearances are characteristic and easily seen. Wash the stomach or other part with distilled water a number of times; filter, and test separate portions for the acids by appropriate tests before mentioned; in many or most cases, no trace is left of the acid.

Hydrocyanic, or prussic acid, (HCN or HCy) was discovered by Scheele in 1782, who named it prussic acid. It is found in cherry laurel, bitter almond, and in the kernels of many stone fruits, also in many plants of the order Rosaceæ. It also exists already formed in the juice of the bitter cassava. It can be made by taking two ounces of ferrocyanide of potassium, ten ounces of dilute sulphuric acid (one of acid to four of water). This is slowly heated in a retort, the receiver containing about one-half pint of water. The following reaction takes place:

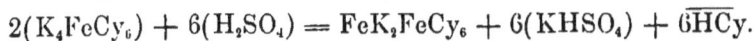

$$2(K_4FeCy_6) + 6(H_2SO_4) = FeK_2FeCy_6 + 6(KHSO_4) + \overline{6HCy}.$$

If pure, the acid should not give a precipitate with sulphuretted hydrogen or barium chloride. The poison is usually in a diluted form, containing from 2 to 20 per cent. of anhydrous acid in solution. The medicinal dose of the diluted (2 per cent.) acid is from two to six drops. The smallest fatal quantity is nine-tenths of a grain of *anhydrous* acid. Coullon has shown that hydrocyanic acid affects *all* animals indiscriminately, and all perish in the same manner. In poisonous doses, the symptoms are not well known, owing to its "lightning action." The patient falls down insensible; there is a cold clammy perspiration, hands clinched, eyes glistening, and pupils dilated. Death takes place, in most cases, with a forcible expiration, which may or may not be accompanied by a shriek. Treatment: if the practitioner gets there in time, which is rarely the case, cold effusions, artificial respiration, and a few drops of ammonia in water, given internally, or brandy; carbonate of ammonia or chloride of lime, held near the mouth or nostrils. The chemical antidote is rarely of any use, because of the rapid action of the poison. A mixture of the proto and persulphates of iron in combination with a little caustic alkali, making harmless potassic ferrocyanide, or if acid is present, the ferric salt would form prussian blue. Emetics and the stomach pump should be used whenever you can. In the post mortem appearances, the lungs, liver, spleen, and kidneys, are invariably gorged with blood; the stomach often exhales the odor of the poison.

TESTS.

(*1*) Nitrate of silver gives with hydrocyanic acid a white amorphous precipitate of the cyanide of silver (AgCN). This precipitate gives with alkaline acetates a red color. The cyanide of silver crystals' are slender prisms, when examined under the microscope — say, about 150 diameters.

(*2*) THE IRON TEST. — Treat a solution of hydrocyanic acid with a little caustic soda, or potash, so as to make an alkaline cyanide (HCN + KHO = KCN + H$_2$O); if we add a few drops of a solution of common green vitrol (sulphate of iron, per and proto salts), it gives a precipitate of prussian blue, and more or less of the proto and sesquioxides of iron; if we add a few drops of hydrochloric or sulphuric acid, these oxides dissolve as chlorides or sulphates of iron, while the prussian blue remains undissolved. (1) In the above, the protosalt converts the cyanide of potash (KCN) into ferrocyanogen, $Fe(CN)_2$, as follows: $2KCN + FeSO_4 = Fe(CN)_2 + K_2SO_4$. (2) The ferrocyanogen combines with the cyanide of potash to form ferrocyanide: $4(KCN) + Fe(CN)_2 = K_4Fe(CN)_6$; or ferrocyanide, and this with an ic salt of iron gives prussian blue $-3(K_4Fe(CN)_6) + 2Fe_2(SO_4)_3 =$

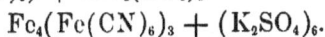

$$Fe_4(Fe(CN)_6)_3 + (K_2SO_4)_6.$$
To make it in one reaction, $18KCN + 3(FeSO_4) + 2Fe_2(SO_4)_3 = 9K_2SO_4 + 3Fe(CN)_2 + 2Fe_2(CN)_6$; and the cyanogen compounds react, and form prussian blue $-3Fe(CN)_2 + 2Fe_2(CN)_6 = Fe_4(Fe(CN)_6)_3$.

(*3*) LIEBIG's TEST, 1847.—Add a few drops of ammonium sulphide to the acid, evaporate gently (not above 100° C.) on a slip of thin glass, a piece of a broken wash bottle will answer, when crystals of sulphocyanide of ammonium will be formed. It is well to notice, if the heat is not carried far enough, the persalt of iron will be precipitated black by the undecomposed sulphide; on the other hand, if the heat is carried too far, the sulphocyanide may be decomposed and lost, a little practice will enable you to get it. A ferric salt of iron with the sulphocyanide gives a blood red color, the sulphocyanide of iron. This test is a very nice one, as you can get the reaction with two or three drops of the solution. It must be noticed that meconic acid (see page 129) gives a red color with the iron, but the red color does not disappear on adding a few drops of a solution of corrosive sublimate, *while, with cyanogen, as above described, it does disappear.* If an acetate is present, the reaction takes place only upon the addition of hydrochloric acid.

(*4*) The red oxide of mercury (HgO) is soluble in solutions of the alkalies *only in the presence of hydrocyanic acid;* now add KHO to the solution, then finely pulverized HgO; if this dissolves, it may be looked upon as a positive test (Fresenius).

(*5*) SCHŒNBEIN'S TEST. — Saturate a sheet of white blotting paper with an alcoholic solution of guaiacum (fifteen grains to one ounce), and dry gently. Dip a slip of the paper into a solution of sulphate of copper (fifteen grains to one ounce), and hold it over a vessel, where the vapor of hydrocyanic acid is given off; then the paper will turn a deep blue. *Ozone gives this reaction, as well as prussic acid.* The quantitative estimation is made by the first test. The bitter taste at the back of the tongue is quite characteristic, also, the odor resembling nitro benzol.

Antimony (Sb^{III-V}, 122) was discovered by Basil Valentine, a monk of Germany, in the fifteenth century. It is found native, but more commonly occurs as stibnite, or gray antimony ore (Sb_2S_3). The oxides are sesquioxide (Sb_2O_3) and antimonic oxide (Sb_2O_5); these are used in painting as a substitute for white lead. It is found native, and known as Valentinite. The oxide forms with potash a number of salts — antimoniate of potash ($K_2OSb_2O_5$, 5aq). Phosphorus, arsenic, and antimony, resemble each other in their oxygen compounds — P_2O_3, As_2O_3, and Sb_2O_3, also, P_2O_5, As_2O_5, and Sb_2O_5; also, their chlorine compounds — PCl_3, $AsCl_3$, and $SbCl_3$, as well as PCl_5, $AsCl_5$, and $SbCl_5$, and in hydrogen compounds — H_3P, H_3As, and H_3Sb; also, in their sulphur compounds — P_2S_3, As_2S_3, and Sb_2S_3, also, P_2S_5, As_2S_5, and Sb_2S_5. Precipitated sulphuret of antimony ($Sb_2S_3 + Sb_2O_3 + 16aq$) is used as an alterative and diaphoretic in combination with calomel and guaiacum, as in Plummer's pills, when mixed with extract of conium or hyoscyamus in the treatment of chronic rheumatism. Kermes' mineral oxysulphuret of antimony ($Sb_2O_3 + 2Sb_2S_5$). There are many other oxides and sulphides, besides most any amount of powders—Jame's powder, oxide of antimony with the phosphate of lime, and Tyon's powder, have about the same composition as the above.

Probably the most important compound of antimony is tartar emetic, a double tartrate of potash and antimony ($KSbOC_4H_4O_6$ + aq). In doses of two to four grains, a powerful emetic; in fractions of a grain, it is a diaphoretic and expectorant. The smallest fatal dose, in the *case of a child*, was two grains. Symptoms: violent and incessant vomiting, metallic taste left in the mouth after taking, intense thirst, and pain in the region of the stomach and abdomen; the pulse is feeble, and intense cardiac depression; the skin, cold and clammy; respiration, laborious; cramps, convulsions, and spasms, often precede death. Treatment: use emetics and the stomach pump to get rid of the poisons, then give most any liquid that contains tannin; as, strong tea, nutgalls, or oak bark decoction, and opium to allay the vomiting. It can be known from most other metallic poisons from the fact, that with ferrocyanide of potassium, it does not give a precipitate. For tests, see comparison of P, As, and Sb, page 41. Antimony and bismuth, when converted into chlorides, and a few drops are added to a large quantity of water, are decomposed into the oxychlorides—$SbCl_3 + H_2O = SbOCl + 2(HCl)$; but bismuth with sulphuretted hydrogen gives a *black* sulphide, while antimony gives an orange red sulphide.

Phosphorus (P, 31) was discovered by Brandt in 1669. It is always found in combination, never free, its chief source being calcium phosphate. It is found in the mineral, animal, and vegetable kingdoms. It should be handled with care, and always cut under water. Its specific gravity is 1.8; its melting point, 110° F. It is soluble in ether, oils, naphtha, and bisulphide of carbon; insoluble in water and alcohol. When taken internally, it enters the circulation, imparts to the breath, urine, and sweat, a garlic odor, and makes these secretions luminous in the dark. It is absorbed by the skin, and after its solution in a fixed oil has been rubbed upon the stomach, all the exhalations are luminous. In small doses, it acts as a stimulant, diuretic, and diaphoretic; in one grain doses, as a corrosive poison, ether and oil, in which it is soluble, hastening its action. The red phosphorus has a specific gravity of 2.14, is not poisonous, and is insoluble in bisulphide of carbon. The vapor of phosphorus causes a necrosis of the jaw bone.

Poisoning has been confined mostly to the use of friction matches, and phosphorus paste for a rat poison. Symptoms: besides those above described, there is a feeling of lassitude, nausea, vomiting, and great thirst; cold perspiration, and feeble and irregular pulse; the abdomen becomes tender to the touch; the extremities, cold; pulse, imperceptible, and other symptoms of collapse. Death may take place in from one to three days. The smallest fatal dose was in the case of a child that died after sucking two matches, the estimated quantity being about one-fiftieth of a grain. Treatment: there is no chemical antidote, you must get it out of the stomach as soon as you can by emetics and the stomach pump; give alkaline drinks, but nothing that will dissolve the poison; as, oil, or fat, etc. Calcined magnesia or chalk, mixed with thick gruel, suspends the poison. Directly opposite physiological action has been ascribed to it by different authors, some contradicting the above statement, and advocating turpentine as an antidote, and also the use of old oil that has absorbed oxygen; hydrogen peroxide (H_2O_2) would be just the thing for this man (Dr. Percy's Prize Essay, 1872). The post mortem appearances are those of a corrosive irritant poison; blood is often found in the bladder, intestines, and pleural cavity. There is a fatty change in the liver, kidneys, glands of the stomach, heart, and muscles.

The tests are described in the comparison, on page 41. In general terms, (1) odor, (2) the property of fuming in the air and shining in the dark, and (3) the fact of its evolving ozone in damp air, are the ordinary tests.

Mercury (Hg^{I-II}, 200). — Discoverer is unknown. Its specific gravity is 13.6. It boils at 360° C., and becomes solid at —40° C. It is widely distributed; it occurs native, and in compounds, its principal ore being a sulphide [(HgS) cinnabar]. It enters into combination with chlorine, bromine, iodine, and sulphur, at ordinary temperatures. Its best solvent is nitric acid. It forms two series of compounds that are very unlike in their properties, as follows:

MERCUROUS SERIES.

Mercurous oxide (Hg_2O), black.

Mercurous sulphide (Hg_2S), black.

Mercurous chloride (HgCl), calomel.

Mercurous iodide (HgI), green.

Mercurous nitrate ($HgNO_3$), mercury in excess.

KHO gives black oxide Hg_2O.

NH_4HO forms a black powder.

$Hg_2Cl_2 + 2NH_4HO = NH_2Hg_2Cl + NH_4Cl + 2H_2O$, a di-mercurosum-chloride.

MERCURIC SERIES.

Mercuric oxide (HgO), yellow to red.

Mercuric sulphide (HgS), cinnabar and vermilion.

Mercuric chloride ($HgCl_2$), corrosive sublimate.

Mercuric iodide (HgI_2), red.

Mercuric nitrate, $Hg(NO_3)_2$, nitric acid in excess.

KHO gives red oxide HgO.

NH_4HO forms a white powder.

$HgCl_2 + 2NH_4HO = NH_2HgCl + NH_4Cl + 2H_2O$, a mercuric ammonium chloride.

TESTS.

The tests are described in the separation of the bases, see pages 35 and 39.

(1) Albumen gives a white somewhat insoluble precipitate. See Millon's reagent, page 90.

(2) Reinsch's test is to be performed the same way as for H_3AsO_3.

(3) To detect the presence of bichloride in calomel, place a drop of HgCl₂ solution on a gold coin; now touch the coin through the drop, with a knife blade. The mercury deposits on the coin as a silvery stain.

(4) The following test for mercurials is very delicate, and well adapted to pill masses: Brighten a strip of copper, and put on it a small piece of the suspected · substance, and moisten with a drop or two of water, made into a paste with the substance; now add a small fragment of KI, and stirring them around, afterwards washing it, a mercurial stain will remain, if Hg is present.

The smallest fatal dose for a grown person is six grains, though some have recovered after taking an ounce. The *ic* salt is more poisonous than the *ous* salt of mercury. All the compounds of mercury are more or less poisonous. Corrosive sublimate ($HgCl_2$) is one of the most active, and, in a medico-legal point of view, the most important. The symptoms are, nausea metallic taste, pain in the stomach, violent vomiting, often containing blood, small pulse, intense thirst, suppression of urine, and stupor. Death is often rapid from collapse. The smallest fatal dose is three grains, children being less susceptible to its action than adults. Death may take place in half an hour, or 'it may not take place for weeks. Treatment: remove the poison by emetics and the stomach pump; give white of egg in milk—albumen in some form; the white of one egg will neutralize four grains of corrosive sublimate.

The cases of chronic mercurial poisoning are generally of long standing, and those engaged in places where mercury is used; as, looking-glass manufactories, and the like. In case of *mercurial tremors,* a dark line may be observed in the gums; the teeth are often brittle, and often sallivation takes place, the tongue and gums becoming red, swollen, and ulcerated. When there is salivation, the parts may commence to slough off.

Arsenic (As^{III-IV}, 75) is found combined with many of the metals; as, Ag, Cu, etc. Its *principal source* is the arsenical pyrites. The substances used in the arts, under the name of arsenic, is really the oxide of arsenic (arsenious anhydride, As_2O_3, and arsenic anhydride, As_2O_5).

Scheele's green is an arsenite of copper, prepared by dissolving white arsenic in a solution of carbonate of potash, and decomposing the arsenite of potash by adding sulphate of copper. Scheele's green ($2CuO, H_2O, As_2O_3$) is used for green paint.

Emerald green is a combination of arsenite and acetate of copper.

Fowler's solution is an arsenite of potassium.

Harle's solution is an arsenite of sodium.

Donovan's solution is an iodide of arsenic added to red iodide of mercury.

Biette's arsenical solution is an arsenite of ammonium.

London purple is an arsenite of lime.

Arsenical soap is made by arsenious acid and camphor, and is used by naturalists for preserving the skins of animals.

It is used in many paints, bug and rat poisons, pyrotechny, medicines, and in making aniline dyes. As commonly sold, it is a white, tasteless solid; sp. gr., 3.7. It is no longer considered a cumulative poison; that is, the continued use of frequent small doses is not believed to possess the power of gradually and silently accumulating in the body, and then suddenly breaking out with dangerous or fatal violence. The symptoms are those of an intense irritant, generally great depression, followed by a "burning pain" in the pit of the stomach, diarrhœa, painful cramps in the legs, and violent vomiting, the vomit being a brown liquid, if mixed with blood, or white, if mixed with the poison; the thirst is intense; the skin, dry and hot; the headache, severe; the pulse, small and rapid; the tongue, dry and furred; breathing, catching; and the mind, generally clear. The symptoms may terminate in many ways—

1. Convulsions, with fits of an epileptic nature.
2. Collapse, with or without pain, and vomiting or diarrhœa.
3. Intense coma.

4. Immediate, as if by shock.

5. It may act like cholera, and deceive the practitioner.

Treatment: give emetics of sulphate of zinc or mustard, afterwards give, as a drink, a mixture of milk and white of eggs. The chemical antidote, due to Bunsen and Berthold, is the hydrated sesquioxide of iron ($Fe_2O_3 3H_2O$). The chemistry of the operation is thought to be as follows: The arsenious acid is oxidized to arsenic acid by the oxygen of the iron, while the iron is reduced to the protoxide ($2Fe_2O_3 + As_2O_3 = 4(FeO) + As_2O_5$); now this protoxide of iron combines with arsenic acid to form an insoluble arsenate of iron, about fourteen parts of the moist, recently precipitated iron being used to one of the arsenic. The hydrated sesquioxide is made by precipitating a ferric salt of iron by an excess of ammonia, straining through muslin, pocket-handkerchief, or anything handy, and washing out the ammonia. It must be borne in mind that the Latin maxim is true in this case at least, "*corpora non agunt nisi soluta*" — a body does not act unless in solution. If the arsenic is not in solution, the hydrated oxide of iron or magnesia is of no use. This statement we verified in the following manner: two small dogs were taken, to each of which eight grains of arsenious acid were given; to one it was given in a solution with milk, and in five minutes the oxide of iron was given, and although symptoms of poisoning were marked, the animal recovered and got well; in the other case, the poison was finely pulverized, and administered dry on beef. In one-half an hour, the first symptoms were shown, when the oxide was administered, but the usual symptoms of arsenical poison went on, and the animal died in six hours. The post mortem appearances: arsenic has the power of preserving the body from decay; the blood is usually fluid, but this is the case with most animals that die a violent death; the organs have the appearance of common irritant poisons.

Christison, on poisons, regards the confirmation by the following reagents as *unimpeachable evidence of the presence of arsenic;* while there are fallacies to any one test taken alone, there is "no

other substance in nature (but arsenic) which produces the same
effects as it with the whole three tests in succession."

TESTS.

(1) Sulphuretted hydrogen in a *slightly acid* solution gives a
lemon yellow precipitate (As_2S_3). It is well to heat
the solution to boiling before passing the H_2S through.
Cadmium gives a yellow precipitate, *insoluble* in
ammonia, but arsenic is soluble in ammonia.

(2) HUME'S TEST (1789) — ammonia nitrate of silver. It
must be made fresh each time it is required, as
follows : Add to a solution of nitrate of silver a weak
solution of ammonia, drop by drop, until the brown
precipitate first produced is *nearly* dissolved, and pour
off the clear solution; this gives with a solution
of arsenious acid a bright yellow precipitate of
arsenite of silver (Ag_3AsO_3). You may find some
trouble to get this precipitate, as the reagent is
decomposed by organic matter.

(3) SCHEELE'S TEST.—The ammonio sulphate of copper is
made by adding to a weak solution of sulphate of
copper, ammonia, drop by drop, until the precipitate
first formed is nearly dissolved; the clear liquid is
now decanted and used. It gives with arsenious acid,
a light apple green precipitate of arsenite of copper
(Scheele's green ($CuHAsO_3)_2$).

Any or all of these precipitates can be afterwards confirmed
by Bloxam's, Marsh's, Reinsch's, or other tests. If an arsenite
be mixed with a large excess of KHO or NaHO in a test tube,
and boiled with Zn or Al, the nascent H evolved combines with
As in arsenites to form AsH_3, which can be tested by a paper
moistened by a silver nitrate, giving a *purple black color*. It does
not act upon Sb compounds, to form SbH_3; this test serves to
distinguish As from Sb (Fleitman's test). Reinsch's test is the
best for complicated organic liquids or mixtures.

Lead (Pb, 207) was known to the ancients. It is not found native, but occurs almost entirely as a sulphide — "galena" (PbS). In small quantities, it is found as a carbonate, cerussite — "white lead ore" ($PbCO_3$), and as a sulphate — "anglesite" ($PbSO_4$). The physical properties of lead are known to every one. Lead rapidly oxidizes from the combined influence of air and water, the water dissolving the lead oxide formed by the action of the air, leaving a clean surface of lead for the further action of the air. This solution of the oxide absorbs carbonic anhydride from the air, and a basic lead carbonate is precipitated ($PbCO_3Pb(OH)_2$). This process can go on continuously. The presence of chlorides, nitrates, nitrites, and ammonia, in the water *promote* this corrosion, while sulphates, phosphates, and carbonates, *hinder* it.

. The following compounds of lead are used in pharmacy — litharge (PbO), minium, or red lead (Pb_3O_4), a union of the proto and di oxides, lead acetate, or sugar of lead ($Pb(C_2H_3O_2)_2$), subacetate of lead, $Pb_2O(C_2H_3O_2)$, lead carbonate, $2(PbCO_3) + Pb(OH)_2$, lead nitrate ($Pb(NO_3)_2$), lead iodide (PbI_2), and lead chloride ($PbCl_2$). It must be noticed that every salt of lead is poisonous, if it is in a condition fitted to be absored by the skin, or mucous membrane, of the stomach. The acetate and carbonate are the most important salts to the toxicologist.

The symptoms of chronic poisoning by lead are well marked. There is pain, with sinking, about the navel, the seat of the colon; constipation, loss of appetite, thirst, fetid odor ·of the breath, and general emaciation, especially the muscles of the arms, and a blueness of the edges of the gums. The effect on the nervous system: there is a numb feeling in the skin, trembling of the arms and legs, a paralysis of the extensor muscles so the hand drops; the body becomes emaciated, the legs œdematous, and the person dies exhausted. In acute poisoning, these symptoms come on rapidly, and with greater intensity. There is invariably constipation from paralysis of the intestinal muscular coat; the urine is scanty and red; violent cramps, cold sweats, paralysis of the lower extremities, and often convulsions and tetanic spasms, come on; the mind is usually clear to the last. The time that death takes place, and the dose of the acetate to produce it, are quite variable — from two drachms to an ounce, and from three to twenty days. Treatment: cause vomiting by sulphate of zinc, warm water, or stomach pump. Give any soluble alkaline or earthy sulphate, sulphate of magnesia is one of the best; it can be given in milk, and mixed with white of eggs. In chronic cases, iodide of potassium is given three times a day in from five to ten grain doses, also, lemonade and sweetened dilute sulphuric acid. But, first and last, great cleanliness is of more importance to the workman than all the doctor's medicine, and at the first indication of lead poisoning, the work must be discontinued, and proper treatment at once adopted; this is the only sure cure for *colica pictorum* (painter's colic). It might be well to state here that the most of the hair washes or hair restorers are solutions of acetate of lead (about four to six grains to an ounce of water) mixed with a little sulphur, and colored and scented with most anything; the greater the lie and humbug, the better it takes, as a rule. The post mortem appearances are quite varied.

TESTS.

(*1*) Sulphuretted hydrogen (H_2S) gives in neutral acid, or alkaline solutions, a black precipitate of PbS; test this by dissolving it in dilute nitric acid, (*a*) gives slender prismatic crystals; (*b*) add to some of this solution ($Pb(NO_3)_2$) iodide of potassium, which gives a yellow lead iodide (PbI_2); (*c*) take another portion of the solution, and add chromate of potash, which gives a bright yellow chromate of lead ($PbCrO_4$); (*d*) dilute sulphuric acid gives a white precipitate of sulphate of lead ($PbSO_4$).

(*2*) Metallic zinc precipitates a solution of lead acetate as metallic lead, that can be dissolved and confirmed by the above tests. H_2S is the most delicate test known for lead.

Zinc (Zn″, 65) is never found native. It was known in the thirteenth century as "spelter," and was described by Paracelsus. It occurs as a carbonate (calamine $ZnCO_3$), sulphide (blende ZnS), and silicate (williamite $2ZnO, SiO_2, H_2O$). In toxicology, the chlorides ($ZnCl_2$), acetate ($Zn(C_2H_3O_2)_2$), and sulphate ($ZnSO_4$), are the most important. The chloride is used by plumbers in soldering, as a flux. It has powerful anticeptic and deodorizing properties; "Burnett's disinfecting fluid" contains from 200 to 250 grains to the ounce. It is a very corrosive poison, and when applied externally, a powerful escharotic. It rapidly coagulates albumen and the delicate tissues of the body. It is a caustic and irritant, producing pain and instant vomiting. The pulse and breathing are accelerated, the voluntary muscles are paralyzed, pupils dilated, coma supervenes, and death comes without a struggle. The sulphate of zinc, as well as the acetate, is rank poison, when administered in large quantities. Treatment: carbonate of soda with milk, white of egg, tea, etc.; opium is given to relieve pain. Post mortem appearances: inflammation of the intestinal tract; the brain and lungs are generally congested.

TESTS.

(*1*) Sulphuretted hydrogen (H_2S) gives in neutral or alkaline solutions (*but not in an acid solution*) a white precipitate (ZnS), the only metal that gives a white sulphide with H_2S. It is best to form a solution of the acetate, an exception to the above.

(*2*) The caustic alkalies give a white $Zn(OH)_2$, soluble in excess of reagent.

(*3*) When heated on charcoal, yellow, while hot, and white, when cold.

(*4*) Take a solution of zinc on platinum foil, and with a piece of magnesium wire, touch the platinum through the solution, when zinc will be deposited on the platinum.

Nearly all the precipitates of zinc are white.

Copper (Cu'', 63.4) is found native, and in compounds, the most important ore being the sulphide of iron and copper ($Cu_2SFe_2S_3$). The carbonate (malachite) and many other compounds are found in rather small quantities. It has a specific gravity of 8.9, and is hard, ductile, malleable, and sonorous. It is a good conductor of heat and electricity, and fuses at 1,996° F. (1,091 ° C.).

Nitric acid dissolves copper to form copper nitrate, liberating nitric oxide $-3Cu + (H_2ON_2O_5)_4 = 3CuON_2O_5 + \overline{N_2O_2} + 4H_2O$. Sulphuric acid dissolves copper to form copper sulphate, and liberating sulphurous anhydride $- Cu + 2(H_2SO_4) = CuSO_4 + \overline{SO_2} + 2H_2O$, while liquid hydrochloric acid acts upon the metal, when finely divided, evolving hydrogen $- Cu + 2HCl = CuCl_2 + \overline{2H}$; when Cu is placed in a solution of chlorides; as, common salt, it becomes coated with a green oxychloride ($CuCl_2$, $3CuO$, $4H_2O$). Copper vessels are therefore dangerous for culinary purposes. Ammonia dissolves copper, but the fixed alkalies have very little action upon it.

It is used in medicine, for coinage, for alloys, for use in the arts; as, sheathing ships, galvanic batteries, and electro plates, etc. In medicine the sulphate ($CuSO_45H_2O$) is one of the most important, in one-fourth to one-half grain doses as a tonic and astringent, and in five grain doses as a powerful emetic. It is also used in gonorrhœa as an injection, in two to eight grains to the fluid ounce. A crystal, polished by trituration on a damp cloth — never scrape it with a knife to modify its shape — is applied as an astringent to inflamed or granulated eyelids, and to ulcerations of the mouth. The anhydrous salt is a test for water in alcoholic solutions. The carbonate has been given in neuralgia. The oxide is used in organic analysis, and takes the place of the carbonate. The nitrate is used as an injection in gonorrhœa and like diseases.

The carbonate is a natural verdigris, while the acetate is an artificial verdigris. Copper salts produce the ordinary symptoms of irritant poisoning. There is a styptic coppery taste, and a burning heat in the throat; vomiting is an early symptom;

the pulse, small and irregular; the body, bathed in perspiration; scanty or entire suppression of the urine, with cramps in the extremities; death is often preceded by convulsions and insensibility. Treatment: cause vomiting by warm water, or use the stomach pump, after which, milk mixed with sugar, or white of egg are given freely; albumen is said to be an antidote. Post mortem appearances are usually confined to the alimentary canal. The inside of the stomach sometimes has a bluish or greenish appearance. The tests are sufficiently described in the separation of the bases; see page 40.

HOW POISONS DESTROY LIFE.

We trace the poison to the circulation, and observe that death is the result; but there is at present no satisfactory theory to account for the fatal effects, or explain *how it operated.* The blood seems to be so changed by the poison as to render it unfitted to perform its proper functions, but the *modus operandi* is as yet a perfect mystery. The chemist, microscopist, nor physiologist, has not been able to throw any light upon the changes produced by the poison in the blood, or in the organs necessary to life.

It has been clearly shown that no substance acts as a poison, until it has been absorbed, and passes through the arterial capilary system. The sooner the poison reaches the blood, the more rapidly does it act; and it depends not so much upon the quantity, as the amount absorbed in a given time. The time for this absorption, under favorable circumstances, is only a few seconds. The fatal effects are produced, when the absorption takes place more rapidly than the elimination. The fatal proportion of poison present in the blood at any one time is infinitesimally small — one-sixteenth of a grain of strychnine has caused the death of a child in four hours. The blood is about one-eighth of the body by weight, and the proportion of the poison by weight, compared with the blood, would be only one part in millions, and the bite of a cobra is yet in a smaller proportion; but this is not all—the blood, urine, saliva, or milk, of an animal poisoned by the cobra, when injected into another animal, will produce death.

The amount of an organic poison, found in the body after death, is the difference between the amount taken and the amount necessary to produce death, or the amount absorbed and eliminated. A person may die from the effects of poison, and no trace of it remain in the body at the time of death. Death takes place from the changes (chemical or otherwise) porduced in the blood by that portion absorbed, and, when the dose is small, it may be all absorbed. There are a number of poisons for which, in the present state of chemical science, there are *no known chemical tests* — a list of them would be out of place in any publication — but in such cases, physiological tests and circumstantial evidence take the place of chemical ones.

DELICATE TESTS FOR ALCOHOL AND CHLOROFORM.

Alcohol.—Take ten cubic centimeters of the clear liquid, add five or six drops of a dilute (10 per cent.) solution of NaHO or KHO, and warmed to about 50° C. A solution of iodide of potassium in iodine is added until the solution becomes yellowish brown; to it is added *cautiously*, a caustic alkali, which decolorizes it; if alcohol is present, *yellow* hexagonal crystals form after a time.

Chloroform.—Add to the liquid to be tested, some alcoholic soda and a little aniline; on gently warming, a *horrible odor* will be given off, due to the formation of benzo-isonitrile (C_7H_5N).

PART VI.

CHAPTER I.

THE BLOOD.

The blood has a bright red color in the arteries, dark purple in the veins, saltish taste, alkaline reaction, and an odor, when warm, resembling the exhalations of the animal from which it was taken; sulphuric acid assists in developing this odor, and augments it when developed. It is about one-eighth the weight of the body, and is composed of two elements—a solution of various substances, the plasma, and formed elements suspended in the plasma, the corpuscles, of which there are two varieties—the red and white, the latter commonly called leucocytes; the relative proportion of the white and red corpuscles is very variable, and is varied by age, sex, period after food, and the vascular territory examined, the average being about one white to 400 of the red. In volume, the plasma exceeds the corpuscles in the proportion of sixty-four to thirty-six, with the following specific gravities: Defibrinated blood, 1.052 to 1.057; plasma, 1.027 to 1.029; leucocytes, 1.070; red corpuscles, 1.088 to 1.100.

The red corpuscles are biconcave bodies, and circular in form; in the human species, nuclei appear at a very early period of life, but subsequently are invisible, unless displayed by artificial means. The size of the red corpuscles vary in different animals; in the human species, in different races, and in individuals of the same race, also, in the time elapsed after the withdrawal of the blood from the vessel. While there are many contradictory views about the red corpuscles, nearly all the

THE BLOOD. 157

authors agree in saying that ninety-five or more out of every 100 are of an uniform size, and that those that vary may be one-third larger or one-third smaller. The average diameter of the human red corpuscle is about .0075 millimeters, or 1-3370 of an inch. The thickness is more variable than the diameters, the average central thickness being about .0017 millimeters. The number of red corpuscles in a cubic millimeter is about 5,000,000. The whole number in an average-sized man has been estimated upwards of 14,000,000,000,000, with an average corpuscular surface of 1,927,000,000 square millimeters. After the blood has been drawn from the vessels, the surfaces and borders of the red corpuscles lose their smooth appearance, the borders become dentated, and the surface beset with little eminences, and they are called crenated. In addition, the red corpuscles become smaller and more spherical. In their normal condition, the red corpuscles contain no cell membrane nor nucleus; in other words, they are homogeneous. They contain an organic nitrogenized principle, coloring matter, and an inorganic principle; the name globuline is given to the former. The coloring matter exists in two conditions — oxyhæmaglobine, in the arteries, and hæmaglobine, in the veins, and the inorganic matters are inseparable, except by incineration. The red corpuscles contain the same organic constituents as the plasma, and besides these, cholesterine, lecithin, and fatty matters.

The plasma has been divided into five classes — inorganic, organic saline, organic non-nitrogenous, excrementitious, and organic nitrogenous. The first class always exists in combination with organic principles, being entirely formed from materials introduced from without the economy — water forms 785 parts in 1,000; the chlorides of potassium, sodium, and ammonium, the sulphates of soda and potash, carbonates of lime, soda, potash, and magnesia, phosphates of lime and magnesia, also, traces of silica, copper, and lead, are sometimes found, potash and soda forming the largest part of this class; it is of definite chemical composition, and crystallizable. The second class is principally formed in the organism, and is present in very small quantities as follows : oleate, magarate, stearate, and lactate of soda and lime.   The third class also exists in small quantities, being formed principally from the food ; the liver acts as an auxiliary in the formation of the saccharine principle and glycogenic matters.   Oline, margarine, stearine, lecithin, glucose, glycogenic materials, and inosite, comprise this class ; they are of organic origin, definite chemical composition, and crystallizable.   The fourth class is formed by the disassimilation of the tissues — carbonic acid, urea, urates of soda, potash, lime, magnesia, and ammonia, sudorates of soda, inosates, oxalates, creatine, leucine, hypozanthine, and chlorestine.   The fifth class receives the materials for its regeneration from the nutritive fluids.   It exists in a condition that is constantly changing ; hence it may be said, that it is endowed wilh vital properties.   To this class belong, plasma, serine, peptones, etc.   This class is of organic origin, indefinite chemical composition, and not crystallizable.

The coagulation of the blood is due to the separation from the plasma, of a body called fibrine, which entangles in its meshes, the corpuscles of the blood, the mechanical interlocking of the corpuscles, by the the threads of fibrine, giving rise to the crassamentum, or blood clot.   The time of coagulation is not the same in all animals, being in the following order : rabbit, sheep, dog, man, ox, and last, the horse, in the horse commencing in five to ten minutes after the blood is drawn.

HOW TO TEST BLOOD STAINS.

There are three methods of identifying blood — (1) by the microscope, (2) by the microspectroscope, and (3) by chemical tests, or reagents.

BY THE MICROSCOPE. — It is found in *man* and all, *mammalia*, except the camel tribe, that the blood corpuscles are circular, and apparently without nuclei ; in the camel, they are *oval* shape. [To measure the size of blood corpuscles, take neutral or alkaline urine, add five grains of corrosive sublimate to every ounce; this precipitates urates; let this stand for six or eight hours, and then pour off the clear fluid, and reduce with water to specific gravity 1.020; this remains clear, and does not allow growths of vegetable matter. It makes a good mixture with blood, the only objection being it bleaches the red corpuscles quickly, and increases their size somewhat].* In birds, reptiles, and fishes, they are generally of a larger size, oval, and nucleated. The observer may be able to state positively that the blood was or was not the blood of a mammal; beyond this, it would be a matter of opinion. Moisten the stain, if recent, with diluted glycerine, sp. gr., 1.025, and after some time, pressing the liquid, examine for corpuscles. In the dog, the constant companion of man, the corpuscles are but little smaller (see Gulliver's tables). In a noted murder trial in Michigan, this was a debated question, as to identifying dog's from human blood. It has been proposed to use one-twenty-fifth and one-fiftieth objective, when the size of the corpuscles in the latter case would be over an inch. We have found as much difference in examining the micro-millimeter scales (four different manufactories' scales were examined) as we did in the blood corpuscles of the dog and man.

Cut out the part of the garment, etc., that contains the stain, and by a thread suspend it in a few drops of distilled water in a watch glass for half an hour; now with a glass rod squeeze out the liquid, and remove the part; this solution is examined for *oxidized hæmoglobine.* Another part of the solution can be reduced by ferrous ammonium sulphate by first adding a little Rochelle salt to keep up the iron. When the *reduced hæmoglobine* is examined by the microspectroscope, it gives a characteristic spectrum. In old stains, it may be hard or impossible to obtain

---

*The following dilute solutions have been employed—a solution of sulphate of soda and distilled water, sp. gr., 1.025. A solution of gum acacia, sp. gr., 1.020; one part added to three parts of a solution of equal parts of sulphate of soda and chloride of soda, so the solution has a sp. gr. of 1.020; mix (Potain's solution). Key's solution (the urine) can be obtained at any time, no cost, and answers every purpose.

the blood corpuscles, or the coloring matter of the blood, when the following *Hæmin* test is of value:

When hæmin, or hæmoglobine, is heated with glacial acetic acid (a few drops) and common salt (the smallest quantity), a hydrochlorate of hæmatine is formed,* which, on evaporation, is deposited in the form of reddish brown prisms — the *hæmin crystals.* These crystals are examined with a power of from 300 to 400 diameters. These crystals can be burned on platinum foil, and show the presence of iron.

DR. DAY'S GUAIACUM RESIN TEST. — Make a fresh solution each time of resin of guaiacum by dissolving 0.5 gram in ten cubic centimeters of alcohol. Moisten the stain with the guaiacum tincture, which should not blue it by itself; now add a small quantity of an ethereal solution of peroxide of hydrogen (a watery solution will do) or ozonic ether. In the presence of blood, the guaiacum is oxidized, and acquires the characteristic bright blue color. This is a good confirmatory test; but it must be remembered that many other substances besides blood; as, gluten, wheaten flour, gum arabic, milk, roots and stems of many kinds, nitric acid, ether, creasote and carbolic acid, mixed with mucus, pus, and saliva, ozone, chlorine, chloride of iron, and many of the metals, give a blue color with guaiacum resin.†

---

* Hoppe-Seyler ascribes to it the formula — $C_{68}H_{70}N_8Fe_2O_{10}2HCl$, while Thudichum says it contains no chlorine. Take seven parts of water and one of glycerine; some recommend a solution of chloral hydrate (one to ten of water).

† This test is the simplest and easiest performed. and is a good confirmatory test. The haemin test takes the next rank in simplicity, requiring only a microscope of comparitively low power (400 diameters), and is very characteristic, especially when burned on platinum foil, and the presence of iron shown. The haemoglobine test requires a microspectroscope, and considerable experience in the use of it, though quite characteristic. All, taken together, may be regarded as positive proof.

CHAPTER II.

SEPARATION OF PLANT BASES.

The following plan for separating the plant bases, or vegeto-alkaloids, is after Dragendorff:

The substance is finely divided, and treated with water and a small quantity of sulphuric acid; this solution is partly neutralized with magnesia, and evaporated to a thick syrup; the residue is treated with alcohol, mixed with dilute sulphuric acid. The alcohol is distilled off, and the aqueous residue filtered, and afterwards shaken up at 40° C. with petroleum ether, which takes up *piperine*, and gives a blood red with strong sulphuric acid, disappearing on the addition of water. The remaining aqueous solution is nearly neutralized with magnesia or ammonia, and is treated with benzine, which takes up caffeine, delphinine,, colchicine, cubebine, digitaline, which is not an alkaloid, and traces of veratrine, physostigmine, and berberine.* The slightly acid liquid is then shaken up with amyl alcohol, which takes up theobromine, and traces of narcotine. aconitine, and atropine.†

The residual aqueous liquid is treated with chloroform, which takes up papaverine and thebaine, with small quantities of narceine, brucine, physostigmine, berberine, narcotine, and

---

*Caffine gives with chlorine water and ammonia the murexide color. Digitaline gives with strong sulphuric acid a red color, and yellow or green when diluted with water. Veratrine gives with strong sulphuric acid a yellow color, changing to crimson and violet. Solutions of colchicine and berberine leave yellow residues on evaporation: the former dissolves in strong sulphuric acid with a dark yellow color, the latter with an olive green color, disappearing on evaporation, and leaving the solution colorless. Colchicine and berberine are separated by iodine, the first, a brown, and the second, green spangles. Physostigmine is not colored by sulphuric acid, and contracts the pupils of the eye, when administered.

† Theobromine may be recognized by its property of gradually dissolving in water, with a yellow color, and changing to blue with ammonia, and giving no color with sulphuric acid.

veratrine.* The liquid, separated from the chloroform extracts, is treated with ammonia, which is added to it under a layer of petroleum ether at 40° C., the vessel being shaken immediately after the liquid has become alkaline. The petroleum-ether dissolves strychnine, brucine, quinine, emetine, veratrine, conine, nicotine, and papaverine.†

The alkaline aqueous solution is next treated at 40° to 50° C. with benzene, and dissolves out quinine, cinchonine, atropine, hyoscyamine, aconitine, physostigmine, and codeine.‡ The bases which may still be present in the alkaline aqueous residue are, morphine, solanine, curarine, and small quantities of narceine and berberine.§

---

*Papaverine gives a blue violet color with strong sulphuric acid. The baine gives a red color with strong sulphuric acid.

†Nicotine and conine may be dissolved out by water. Neutralize with sulphuric acid, and precipitate nicotine with potassic cadmic iodide *crystals*, conine *amorphous*. The other bases, freed from petroleum ether, dried, etc., are treated with ether, which takes up quinine, emetine, papaverine, and veratrine. Dissolve in the smallest quantity of sulphuric acid; now add sodium carbonate; this precipitates quinine, emetine, and papaverine. Strychnine and brucine remain in solution, and can be separated by means of alcohol, strychnine being quite insoluble in alcohol.

‡Evaporate to dryness, and treat with ether; chinchonine remains behind. On evaporating the ethereal extract, and dissolving the residue in very dilute sulphuric acid, and mixing the solution with a slight excess of ammonia, quinine and aconitine are precipitated, and atropine, hyoscyamine, and codeine remain in solution; aconitine and quinine are separated by dissolving in a small quantity of hydrochloric acid, and adding platinic chloride, which precipitates quinine; now remove the platinic chloride in the solution by hydrosulphuric acid, and dissolve out aconitine by chloroform. Atropine can be told from hyoscyamine by treating the former with potassium bichromate and sulphuric acid, by the odor, also, by the action on the pupil of the eye.

§From this mixture, morphine and solanine, besides small quantities of narceine, are separated by acidulating with sulphuric acid, and heating to 50 or 60 degrees C., and covering the surface of the liquid with a layer of amyl alcohol, and then adding an excess of ammonia, and agitating. The morphine crystallizes out from the amyl-alcoholic solution, while the solanine gelatinizes as the liquid cools. Curarine is characterized by giving with sulphuric acid and potassium bichromate, a color similar to strychnine. See pages 125 and 126.

CHAPTER III.

LAWS OF CHEMICAL INCOMPATIBILITY.

Substances are said to be incompatible when their combination gives rise to chemical changes, a new compound being formed which is either inert, or possessed of some different properties. But the student must bear in mind the difference between chemical and therapeutical inertness. Living beings can dissolve, appropriate, and circulate in their fluids, substances which, under ordinary circumstances, are the most intractable and insoluble. A number of the mineral salts of platinum, gold, copper, lead, tin, and zinc, are sometimes precipitated with albuminous substances, forming ordinarily insoluble compounds, but are soluble in the liquids of the alimentary canal, and may be in a condition suitable for medicinal action. [The antidote for corrosive sublimate ($HgCl_2$) is albumen; the antidote for arsenious acid (As_2O_3) is hydrated peroxide of iron; these form insoluble compounds; but it is thought that they are not perfectly insoluble, but that their absorption is diminished, and their immediate effects destroyed. In the third edition of Taylor on Poisons, pages 294 and 305: "The oxide of iron appears to have no more effect on *solid* arsenic than so much brick dust." It is further substantiated that the urine often contains the antidote and poison for five or six days after the cure is said to have been effected].

BERTHOLLET'S LAW.—1. Two salts in solution may form by the interchange of their acids and bases, two insoluble salts, which are precipitated. Sometimes in the above reaction, a *soluble* and an *insoluble salt* may be formed, and the *insoluble* may be precipitated, or may form with the other a *double soluble salt*, and no precipitate occurs.

21

2. In mixing any salt combined with a *weak acid*, and afterwards adding a *stronger acid*, a salt of the stronger acid is formed, and may or may not be precipitated, depending upon the solution.

3. When *two bases* are used, have the *acids alike;* do not mix sulphate of morphine and acetate of lead, but the *acetates of both*.

4. Metallic oxides combine with acids to from salts that may or may not act like either.

5. Alkalies, in contact with the salts of the metals proper, or with the alkaloids, decompose them, precipitating their bases.

6. Vegetable substances, containing tannic or gallic acids, precipitate albumen, vegetable alkaloids, and most of the metallic oxides, and form inky solutions with salts of iron.

7. Glucosides should not be used with free acids or emulsions.

8. The following substances are best given alone — hydrocyanic acid, nitro-hydrochloric acid, antimony tartrate, liquor calcis, and potassæ, the chloride and nitrate of iron, tincture of iodine, the bromide, acetate, iodide, and permanganate of potash, the poisonous metals combined with acetic acid, and, in general, the alkaloids.

CHAPTER IV.

MAN.

ANALYSIS OF A MAN FIVE FEET, EIGHT INCHES HIGH, AND WEIGHING 154 POUNDS.

	lbs.	oz	grs.
1. Water is found in every tissue and secretion....	109	0	0
2. Fibrine forms solid materials of muscle, flesh, and blood	15	10	0
3. Phosphates of lime in all tissues and liquids, bones and teeth................................	8	12	0
4. Fat (a mixture of three compounds) distributed throughout the body.......................	4	8	0
5. Ossein (framework of bones and connective tissues) yields gelatine when boiled............	4	7	350
6. Keratin (with other N compounds is found in the skin, epidermis, hair, and nails).............	4	2	0
7. Cartilagin (N compounds constituents of cartilage) resembles the ossein of the bones...........	1	8	0
8. Hæmoglobin (N substance containing Fe)........	1	8	0
9. Albumen (soluble N substance found in chyle, lymp, blood, and muscle)....................	1	1	0
10. Calcium carbonate ($CaCO_3$) found in bones......	1	0	350
11. Kephalin (with myeline, cerebrine, and in brain, nerves)	0	13	0
12. Fluoride of calcium (in bone and teeth).........	0	7	175
13. Phosphate of magnesia (chiefly in bones and teeth)	0	7	0
14. Common salt (NaCl) everywhere................	0	7	0
15. Cholesterin, inosite, and glycogen (in brain, muscle, and liver)	0	3	0
16. Sulphates, phosphates, and organic salts of sodium	0	2	107
17. " " " chlorides of potassium	0	1	300 · ·
18. Silica (in hair, skin, and bone)................	0	0	30
	154	00	000

There are found in the body, sixteen elements, seven metals, and nine non-metals—C, H, N, O, S, P, Cl, F, Si, Na, K, Li, Ca, Mg, Fe, Mn, and in some cases traces of Cu and Pb, probably accidental.

Water contains H and O.

Fibrin, albumen, ossein, keratin, and cartilagin, contain C, H, O, N, and S.

Hæmoglobin contains all the above, and Fe.

Kephalin and myelin contain C, H, N, P, and O.

Cerebrin and kreatin contain C, H, N, and O.

Fat, cholesterine, inosite, and glycogen, contain C, H, and O.

Phosphate of lime contains Ca, P, and O.

Carbonate of lime contains Ca, C, and O.

Fluoride of lime contains Ca and F.

Phosphate of magnesia contains Mg, P, and O.

Common salt contains Na and Cl.

Sulphates contain different metals with S and O.

Silica contains Si and O.

DAILY SUPPLY.

	Lbs.	Oz.	Grs.	Lbs.	Oz.	Grs.
O in air breathed	1	10	115			
O in starch, fat, etc.	0	7	370			
				2	2	47
C in fat, starch, etc.				0	9	400
H " " "				0	1	170
N in albuminoids				0	0	291
NaCl				0	0	325
Phosphates and potash salts				0	0	170
Water				5	8	320
				8	7	410

DAILY WASTE.

	Lbs.	Oz.	Grs.	Lbs.	Oz.	Grs.
O in CO₂ given out by the lungs	1	7	325			
O " " " skin	0	0	111			
O in organic matter given out by the kidneys and intestines	0	0	357			
O in H₂O formed in the body	0	9	130	2	2	47
C in CO₂ given out by the lungs	0	8	320			
C " " " skin	0	0	40			
C in organic matter given out by the kidneys	0	0	170			
C in organic matter given out by the intestines	0	0	308	0	9	400
H in H₂O formed by the lungs and skin	0	1	70			
H in organic compounds that pass off by intestines and kidneys	0	0	100	0	1	170
N in urea given out by kidneys	0	0	245			
N in waste given out by intestines	0	0	46	0	0	291
NaCl given out by skin	0	0	10			
NaCl " " kidneys	0	0	315	0	0	325
Phosphates and potash salts pass off chiefly by the kidneys				0	0	170
H₂O taken in as such and given out by the lungs, skin, kidneys, and intestines, in addition to that formed in the body				5	8	320
				8	7	410

In the language of my friend and teacher, Prof. Norton, "The apparatus is a living machine, governed by an immortal soul, working for itself and others. The machine finally wears out, leaving about seven pounds of lime salts to the hundred and a memory deathless for a chosen few, weak and fugitive for the great majority. Nevertheless, life is worth living, if we have the approval of a good conscience."

PART VII.

GENERAL STOICHIOMETRY.

USEFUL CONSTANTS.

I.

1 micromillimeter	= 1-25,000 of an inch (microscopic unit).
1 gram	= 15.4 grains.
1 kilogram	= 2.2 pounds.
29.57 cubic centimeters	= 1 fluid ounce.
1 liter	= 61 cubic inches, or 2.1 pints.
1 meter	= 39.37 inches, or 100 centimeters.
1 inch	= 2.5 centimeters, or 25 millimeters.
29.92 inches	= 760 millimeters (barometer).

One hundred cubic inches of air weigh 31 grains, and air is 14.43 times heavier than hydrogen. Water at 0° C. (32° F.) is 11,160 times heavier than hydrogen. One grain of hydrogen has a volume of 46.73 cubic inches.

NOTICE. — That (1) 11.19 liters of any simple gas; as, O, H, Cl, etc., weigh as many grams as its atomic weight:* (2) that the density of any aeriform body is one-half of its molecular weight; (3) when gases combine, the product is two gaseous volumes; as, two volumes of hydrogen combine with one volume of oxygen to make *two volumes* of water (in a like manner, three of hydrogen combine with one of nitrogen to form two volumes

---

*Owing to the fact, that mercury, zinc, and cadmium, have each one atom in a molecule, their vapor densities are one-half their atomic weight, and their atomic volumes double that of hydrogen; therefore, it will take 22.4 liters of the vapor of mercury to weigh 200 grams. Also phosphorus and arsenic have each four atoms in a molecule. Their vapor density is double their atomic weight, and their atomic volumes one-half that of hydrogen, and they require but 5.6 liters to weigh as many grams as their atomic weight.

of ammonia); (4) the "crith" is the weight of one liter of hydrogen at 0° C., and 760 m. m. barometer = .0896 grams; (5) the mechanical equivalent of heat is 772 foot pounds for 1° F., or 1,390 foot pounds for 1° C. — 423.6 kilogram meters.

II. Conversion of thermometric scales (C., Centigrade; F., Fahrenheit; R., Reaumur).

To reduce F. to C., 5-9 (F.° — 32) = ° C.
" " C. to F., 9-5 C.° + 32 = ° F.
" " R. to F., 9-4 R.° + 32 = ° F.
" " F. to R., 4-9 (F.° — 32) = ° R.
" " C. to R., 4-5 C.° = ° R.
" " R. to C., 5-4 R.° = ° C.

At what point are the numbers on the scales of Fahrenheit and Celsius identical?

Let x equal the number:

$$32° + 9\text{-}5 \text{ of } x° C. = x° F.$$
$$160° + 9x = 5x$$
$$160° = -4x$$
$$-40° = x.$$

III. Gases expand 1-273 part of their volume at 0° C. for every increase in temperature of 1° C.; the contraction follows the same law. Expressed decimally, 1-273 = .003665, and is called the coefficient of the expansion of gases. The volume of a gas varies *directly* as its absolute temperature, and *inversely* as to the pressure to which it is subjected.

Five hundred cubic centimeters of a gas at 5° C. is heated until it becomes 700 cubic centimeters. Through what number of degrees C. has the gas been heated?

Let x equal number of degrees C., through which the gas has been heated,

$$500 : 700 :: 273 + 5 : 273 + x$$
$$500 (273 + x) = 700 \times 278$$
$$x = 81° C., \text{ nearly.}$$

A liter of air is measured at 0° C. and 760 mm. What volume will it occupy at 720 mm. and 18° C.?

$$\left. \begin{array}{l} 273 : 273 + 18 \\ 720 : 760 \end{array} \right\} :: 1,000 : x. \quad \text{Solve for } x.$$

IV. The specific gravity of a body is its weight compared with the weight of an equal volume of the standard. Hydrogen and air are the standards for gases, water for liquids and solids. In the above normal condition of temperature and pressure are understood

W equals weight in air; w equals loss in water.

$\therefore$ Specific gravity equals $\dfrac{W}{w}$ (for solids).

A body weighs. in the air, 450 grams; in water, 240 grams. Require the specific gravity :

$$\begin{array}{r} 450 \\ 240 \\ \hline 210 \end{array} \qquad \frac{450}{210} = 2.2 \text{ (nearly)}.$$

For getting the specific gravity of a small quantity of liquid see urinary analysis. page 82.

Notice.— (1) The specific gravity of a simple gas is its atomic weight as compared with hydrogen; if this is divided by 14.43, it will give the specific gravity, as compared with air. Oxygen is 16 times heavier than hydrogen, or $\frac{16}{14.34} = 1.1$ times heavier than air. (2) The specific gravity of a compound gas, as compared with hydrogen, is one-half its molecular weight —

$$CO_2. \quad \begin{array}{r} C = 12 \\ O_2 = 32 \\ \hline 2\,)\,44 \\ \hline 22 \end{array}$$

times heavier than hydrogen, or 1.5 times heavier than air —

$$\begin{array}{ll} NH_3 = N = 14 & CH_4 = C = 12 \\ H_3 = 3 & H_4 = 4 \\ 2\overline{)17} & 2\overline{)16} \\ 8.5 & 8 \end{array}$$

and so on.*

V. To calculate the percentage composition of a compound from its formula: ·(1) Calculate the percentage composition of potassium nitrate (KNO_3)?

---

*1. A piece of cork weighs in the air eighty-two grams. When weighed in the air with a piece of metallic tin, they weigh 991 grams; when weighed in the water, they weigh 335 grams. The tin itself loses 124 grams. What is the specific gravity of the cork?

991 — 355 = 636, — 124 = 512; 82 divided by 512 = .16, the spcific gravity of cork.

2. Hiero, king of Syracuse, gave his goldsmith fourteen pounds of gold, and three and one-half pounds of silver, to make a crown. Suspecting that the gold had not been all used, he requested Archimedes to find how much had been abstracted, the specific gravity of gold being nineteen and one-fourth; of silver, ten and one-half; and of the crown, fourteen and five-eighths. Problems of this class properly belong to alligation.

$$14\tfrac{5}{8} \left\{ \begin{array}{c} 10\tfrac{1}{2} \\ 19\tfrac{1}{4} \end{array} \right\} \begin{array}{c} 4\tfrac{5}{8} \\ 4\tfrac{1}{8} \end{array} \text{ or } \begin{array}{c} 37 \\ 33 \end{array} \text{ in bulk} = \begin{array}{c} 37 \\ 33 \end{array} \text{ multiplied by } \begin{array}{c} 19\tfrac{1}{4} \\ 10\tfrac{1}{2} \end{array}, \text{ or } \begin{array}{c} 37 \\ 60\tfrac{1}{2} \end{array} \text{ or } \begin{array}{c} 74 \\ 121 \end{array} \text{ or } \begin{array}{c} \text{silver.} \\ \text{gold.} \end{array}$$

Now, by weight,

$$74 + 121 = 195; \ \frac{121}{195} \text{ of } 17\tfrac{1}{2} \text{ pounds } (14 + 3\tfrac{1}{2}) = 10\frac{67}{78} \text{ pounds of gold.}$$

$$\frac{74}{195} \text{ of } 17\tfrac{1}{2} \text{ pounds} = 6\frac{25}{39} \text{ pounds of silver.} \quad 14 - 10\frac{67}{78} = 3\frac{11}{78} \text{ pounds of gold abstracted.}$$

The general solution of this problem is as follows:

Let M be the mass of the body, and n its specific gravity. Let H be the mass of the heavier substance, and t its specific gravity. Let L be the mass of the lighter substance, and l its specific gravity. Then, $M = H + L$. Since the volume of a substance equals its mass, divided by its specific gravity,

$$\frac{M}{n} = \frac{H}{t} + \frac{L}{l}$$

From these two equations, it is found that

$$H = M\frac{t}{n}\left(\frac{n-l}{t-l}\right) \qquad \therefore L = M\frac{l}{n}\left(\frac{t-n}{t-l}\right)$$

The specific gravity of the mass can be determined the usual way—*see urine;* the specific gravity of the components can be found by tables, or from fragments of the body, when the proportion of the ingredients may be found by the above formulas.

22

$$K = 39.$$
$$N = 14.$$
$$O_3 = 48.$$
$$\overline{101.}$$

$$O = \frac{48 \times 100}{101}; \quad N = \frac{14 \times 100}{101}; \quad K = \frac{39 \times 100}{101}.$$

(2) How much mercury is contained in 125 pounds of an ore of which 75 per cent. is mercuric sulphide (HgS)?

Multiply 125 by $\frac{75}{100}$ = 94, pounds, nearly, of mercuric sulphide.

$$Hg = 200$$
$$S = 32$$
$$\overline{232}$$

$$Hg = \frac{200}{232} \text{ of } 94$$
$$S = \frac{32}{232} \text{ of } 94$$

VI. To calculate the amount of material required to produce a given weight of any substance; or the quantity of the substance produced by the decomposition of a known weight of the material.

1. We want fifty pounds of oxygen. How many pounds of potassium chlorate must we take?

$$KClO_3 = KCl + \overline{O_3}$$

$$K = 39$$
$$Cl = 35.5$$
$$O_3 = 48$$
$$\overline{122.5}$$

$$O = \frac{48}{122.5} \qquad 48 : 122.5 :: 50 : x$$
$$\therefore x = 127.7 \text{ lbs.}$$

From IV we can easily get its volume, if it is required.

2. The silver is to be precipitated from 100 grams of silver nitrate by means of metallic zinc. How much zinc will be required?

(108) (14) (16)   (65)

$$2AgNO_3 + Zn = Zn(NO_3)_2 + Ag_2$$

$2(108 + 14 + 48)$ = 340 parts of silver nitrate, require sixty-five parts of zinc.

$\frac{65}{340}$ of 100 = 19 (nearly) grams.

VII. The combinations and decompositions of bodies in the gaseous form (see I, notice 3).

1. How many cubic feet of oxygen are required to consume completely one cubic foot of marsh gas, and how many cubic feet of carbonic anhydride and of water will be formed?

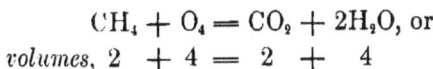

$$CH_4 + O_4 = CO_2 + 2H_2O, \text{ or}$$
$$\text{volumes, } 2 + 4 = 2 + 4$$

[H_2O itself is two volumes, but it requires two of water].

∴Two cubic feet of marsh gas require four cubic feet of oxygen, and there are formed two cubic feet of carbonic anhydride and four cubic feet of watery vapor; now one cubic foot of marsh gas requires two cubic feet of oxygen, and one cubic foot of carbonic anhydride and two cubic feet of watery vapor are formed.

2. X volumes of ammonia are decomposed by chlorine. How many volumes of chlorine are required?

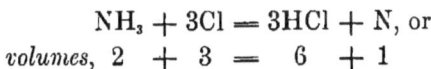

$$NH_3 + 3Cl = 3HCl + N, \text{ or}$$
$$\text{volumes, } 2 + 3 = 6 + 1$$

You can see that two vols. of NH_3 require three vols. of Cl.

∴ one " " " $1\frac{1}{2}$ " "
∴ x " " " $1\frac{1}{2}$ x "

VIII. How the atomic weight of an element is obtained.

1. If it is a gas, by comparing the same volume under like conditions of temperature and pressure with hydrogen.

2. If a metal, the product of its specific heat by its atomic weight is a constant quantity, about 6.34. The specific heat of a body is the amount of heat required to raise a unit weight of the substance from 0° C. to 1° C., as compared with an equal weight of water.

(*a*) In case of the gaseous elements, by simple determination of their densities.

(*b*) In a general way, by making an analysis of their compounds — if possible, gaseous compounds — and carefully comparing the results.

(c) Sometimes a formula is assumed. The formulæ of compounds used are very necessary. Much assistance is derived from — (1) Mitscherlich's law of isomorphism; (2) law of specific heat; (3) by substituting in portions, as from H_2O, KHO, K_2O, etc.

(d) In gases; if the density is doubled, it equals its molecular weight.

1. Stas found, after adding 7.25682 grams of potassium chloride to 10.51995 grams of silver, dissolved in nitric acid, that .0194 grams of silver remained in solution. Calculate from these data the atomic weight of potassium (the other atomic weights supposed to be known).

$$10.51995 - .0194 = 10.50055,$$ the amount of silver used up.
At. wt. Ag : at. wt. Cl :: wt. Ag : wt. Cl
    108   :   35.5   :: 10.50055 :   x, or 3.45157.
$$7.25682 - 3.45157 = $$ wt. of K, or 3.80525.
wt. Cl : wt. K :: at. wt. of Cl : at. wt. K.
3.45157 : 3.80525 ::   35.5   : x; x = 39.1

2. Erdmann and Marchand obtained 109.6308 grams of mercury from 118.3938 grams of the red oxide. Calculate the atomic weight of mercury, supposing oxygen to be known.

$$HgO - Hg = O.$$
$$118.3938 - 109.6308 = 8.7630,$$ the weight of the oxygen.
Wt. of O : at. wt. of O :: wt. Hg : at. wt. Hg.
8.7630 : 16   :: 109.6308 :   x

Solve for x, when you find x equals 200, nearly.

IX. The percentage composition of a body being given, required its empirical formula.

1. A substance has been found to contain in 100 parts:

Hydrogen equals 2.04 ÷ 1 = 2. ⎫
Sulphur   "   32.65 ÷ 32 = 1. ⎬ = H_2SO_4
Oxygen   "   65.31 ÷ 16 = 4. ⎭
_____
100.00

Rule: Divide the percentage by the atomic weight, reduce the quotients to their simplest relation in whole numbers.

2. Potassium equals $28.73 \div 39 = .73 = 1$
   Hydrogen " $.73 \div 1 = .73 = 1$
   Sulphur " $23.52 \div 32 = .73 = 1$
   Oxygen " $47.02 \div 16 = 2.93 = 4$ $\left.\rule{0pt}{2em}\right\}$ $= KHSO_4$

   $\overline{100.00}$

X. Thermal units.

1. Five kilograms of water have to be raised through 10° C. How much charcoal (calling it pure coal) would it be necessary to burn to do this?

By consulting works on chemistry or physics, you will find that coal, in burning, develops 8080 units of heat.

$8080 : 50 :: 1 : x$, when $x = .00618$ kilograms or 6.18 grams.

2. 1,120 pounds of iron ore has to be raised from the bottom to the top of a shaft 1,000 feet deep. What weight of charcoal would develope, during its combustion, force enough to do this?

1120 multiplied by $1000 = 1,120,000$ foot pounds required. Now, one pound of coal, in its combustion, developes 8080 units of heat; but the mechanical work which this heat is capable of doing is—

8080 multiplied by 1390, or 11,231,200. foot pounds.

$\therefore 11,231,200 : 1,120,000 :: 1 : x. \quad x = .09$ pounds.

3. A piece of zinc falls from a height of 1,000 feet; to what temperature, Centigrade, will the zinc be raised by the arrest of motion?

When water falls from a height, every 1,390 feet fallen generates 1° C. (see note I.)

$1390 : 1000 :: 1 : x. \quad x =$ temperature; to which water would be raised by a fall of 1,000 feet.

$x = .769°$ C.

Sp. heat of zinc : sp. heat of H_2O :: temp. of H_2O : temp. of zinc.

Then, $.0927 \quad : \quad 1 \quad :: \quad .769 \quad : \quad x$

Therefore, $x = 8.3°$ C., nearly.

Students have experienced some dfficulty in solving problems like the following, when the bye products are of two different kinds and vary in amount.

How much $KHSO_4$ and K_2SO_4, HNO_3 will be formed when you use $120KNO_3 + 94H_2SO_4$ in making nitric acid ?

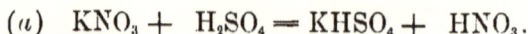

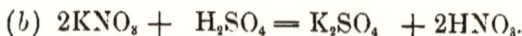

(a) $KNO_3 + H_2SO_4 = KHSO_4 + HNO_3$.

(b) $2KNO_3 + H_2SO_4 = K_2SO_4 + 2HNO_3$.

(c) $3KNO_3 + 2H_2SO_4 = K_2SO_4 + KHSO_4 + 3HNO_3$.

NOTICE. (1) You will have as many parts of nitric acid as you take of potassic nitrate.

(2) The bisulphate ($KHSO_4$) is *first formed*.

(3) If you subtract the number of parts of sulphuric acid taken from the number of parts of potassic nitrate, it will give you the number of parts of normal sulphate (K_2SO_4) formed.

(4) If from the number of parts of sulphuric acid taken, you subtract the number of normal sulphate, you get the number of bisulphate ($KHSO_4$) formed. The answer can now be written out, as follows :

$$120KNO_3 + 94H_2SO_4 = 120HNO_3 + 26K_2SO_4 + 68KHSO_4.$$

The statement is general, and can be put in algebraic language if required. We have solved all these problems for this reason : After taking a number of laboratory classes in the University through Thorpe's Chemical Problems, we have found that the students would get a better knowledge of the subject by taking a special case and working it out than any amount of general rules and statements.

INDEX.

PAGE.

Acids—definition of... .19-20
Acid tests—
  acetic, $C_2H_4O_2$ 1-76
  aconitine, $C_{30}H_{47}NO_7$ ... 118
  arsenic, As_2O_5 42
  arsenous, As_2O_3 42
  benzoic, $C_7H_6O_2$ 74
  boric, HBO_2 66
  carbonic, H_2CO_3 67
  chloric, $HClO_3$ 72
  chromic, H_2CrO_4 9
  citric, $H_3C_6H_5O_7$ 74
  ferricyanic, H_3FeCN_6 .. 75
  ferrocyanic, H_4FeCN_6 . 75
  formic, CH_2O_2 76
  gallic 108
  hydriodic, HI 69
  hydrobromic, HBr 69
  hydrocyanic, HCN 70
  hydrochloric, HCl 69
  hydrosulphuric, H_2S ... 72
  hydrofluosilicic, $(HF)_2$-
    SiF_4 64
  hypochlorous, $HClO$.... 71
  iodic, I_2O_5 68
  nitric, HNO_3 72
  nitrous, HNO_2 71
  nitro hydrochloric, HNO_3
    $+ 3HCl$ 69
  nitro phenic, C_6H_3-
    $(NO_2)_3O$ 66
  oxalic, $H_2C_2O_4$ 66
  perchloric, $HClO_4$ 73
  phosphoric, H_3PO_4 ... 66
  silicic, H_4SiO_4 67
  succinic, $C_4H_6O_4$ 75
  sulphuric, H_2SO_4 64

PAGE.

Acid tests (continued) —
  sulphurous, H_2SO_3.... 67
  sulphanilic 108
  tartaric, $C_4H_6O_6$ 73
  uric, $C_5H_4N_4O_3$ 85
Acid salt................. 20
Albumen, $C_{72}H_{118}N_{18}O_{22}S$. 69
Alcohol, C_2H_6O 4
Alkalies................ .. 55
Alkaline Earths.......... 53
Alkaloids 118
Aluminum (Al) tests...... 50
Alloys 60
Alloxantine, $C_8H_4N_4O_7$ 86
Amalgam 60
Amidogen, NH_2......... 85
Ammonium 55
  carbonate $(NH_4)_2CO_3$. 4
  chloride, NH_4Cl..... . 4
  hydrate, NH_4HO 5
  molybdate $(NH_4)_2MoO_4$ 5
  sulphide $(NH_4)_2S$ 5
  oxalate $(NH_4)_2C_2O_4$.. 5
Analysis 19-23
  of man........ 165
  daily supply........... 166
  daily waste............ 167
Anhydrides — acids and
  bases 18
Antimony (Sb), tests 43
Aqua regia.............. 69
Aricine, $C_{23}H_{26}N_2O_4$ 118
Arsenetted hydrogen,
  H_3As............. 42
Arsenic 18
  ic and ous 41
Artiads 21

PAGE.

Ash—
  milk 113
  cheese 115
Atom 18
Atomic weight... 21
Atomicity 21

Barium, Ba.. 56
  carbonate, $BaCO_3$ 5
  chloride, $BaCl_2$ 5
  hydrate, $Ba(OH)_2$ 6
  nitrate, $Ba(NO_3)_2$ 6
Bases 19
Beads of borax, or micro-
  cosmic salt............ 29
Bell metal... 60
Bennett, Dr............. 120
Benzoic acid, $C_7H_6O_2$... 74
Beryllium 47
Bettendorff........ 42
Bile.................... 95
Binary compounds....... 19
Bismuth, Bi........ 39
Blood 156
  cogulation.... ... 158
  plasma.............. ... 157
  red corpuscles......... 157
  tests 158
Bloxam........... 42
Boettger's tests.......... 93
Boracic acid, HBO_2... 66
Brass 60
Britannia metal......... 60
Bromine, Br............. 3
Bronze 60
Brucine, $C_{23}H_{26}N_2O_4$ ... 118
Butter 112

Cadmium, Cd........... 40
Cæsium, Cs.......... 59
Caffine................ 118
Calcium, Ca........... 57

PAGE.

Calcium (continued)—
  carbonate, $CaCO_3$ 56
  chloride, $CaCl_2$ 6
  hydrate, $Ca(OH)_2$ 6
  phosphate, $Ca_3(PO_4)_2$: 47
  sulphate, $CaSO_4$ 6
Calculi................ 98
Calomel, HgCl.........8–145
Carbamide, urea......... 85
Carbonic acid, H_2CO_3... 67
Carbondisulphide, CS_2... 6
Caseine 112
Celestine, $SrSO_4$ 55
Charcoal—bases heated on 25
Cheese 114
Chemistry definition...... 18
Chloric acid, $HClO_3$ 72
Chlorides, urine 90
Chlorine water 4–130
Chromium, Cr........... 49
Chromic acid, H_2CrO_4... 9
Citric acid, $H_3C_6H_5O_7$.. 74
Cinnabar, HgS.......... 144
Cobalt, Co 50
  nitrate, $Co(NO_3)_2$ 6
Codamine 119
Codeine 118
Colchicine............. 118
Color—
  beads, etc............. 29
  films 30
  oxides 26
Combustion, wet......... 105
Conine................. 118
Copper...... 40
  sulphate, $CuSO_4$ 6
Corrosive sublimate, $HgCl_2$ 145
Cystine 81

Daily quantity of urine.... 78
Daturine 118
Davy's test........... 42
Delphinine 118
Dibasic acid........... 20

PAGE.

Directions—
  how to keep a note-book. 22
  for analysis in dry way.. 31
Distillation, water 104
Donovon's solution...... 147
Double salt............ .. 20
Dry way, tests........... 23

Electrolysis, Mn.......... 57
Element 18
Elements—table of....... 21
Elevation—effect on boil-
  ing water.............. 101
Emetine 118
Ether ($C_2H_5)_2O$ 7
Expansion of water....... 100

Fat—
  cheese 114
  milk 113
Fehling's solution........ 92
Ferrous sulphate, $FeSO_4$. 7
  ammonium sulphate, Fe-
  $(NH_4)_22(SO_4)$ 105
  chloride, Fe_2Cl_6....... 7
Film tests................30-31
Flame reaction........... 27
Fleitman's test........... 42
Fluoride of silicon........ 2
Flux 28
Formula 20
Fowler's solution........ 147

Gallic acid.............. 108
Gas—colored and colorless 64
Gmelin's test............ 95
Gold chloride, $AuCl_3$... . 7
Groups—
  alkalies 58
  alkaline earths 52
  iron.................... 46

PAGE.

Groups (continued)—
  lead and arsenic 36
  silver 32
  Gun metal............... 60
  Gysum, $CaSO_4$.......... 55

Halogen group 4
Hall's test 126
Hardness of water........ 106
Heat substance—
  in glass tube........... 24
  flame blow pipe on C... 25
  with Na_2CO_3 on C..... 27
Heller's test............. 95
Hesse's test.............. 131
Huefner's test............ 84
Hydrates................. 101
Hydriodic acid 69
Hydrobromic acid 69
Hydrocyanic acid........ 70
Hydrosulphuric acid 72
Hydrofuosilicic acid 64
Hydrogen peroxide....... 50
Hyoscamine............. 118
Hypochlorous acid 71
Hypozanthine 119

Incompatibles 163
Iodic acid 68
Iodine, I..............4-65-69
Incrustation and coating .. 30
Inorganic constituents of
  urine........ 80
Iron, Fe............... 49

Jervine 118

Kreatine 80-119

Lanthanium, La.......... 47
Laudanosine 119

(179)

PAGE.

Lead, Pb.... 30-33-36-108
  acetate, Pb($C_2H_3O_2$)$_2$ 7
  alloys................... 60
  chloride, PbCl$_2$....   33
  chromate, PbCrO$_4$ 34
  nitrate, Pb(NO$_3$)$_2$.... 34
  sulphide, PbS.......... 34
Leptandrine............. 118
Lithium, Li............. 58
Liquor potassa, KHO.... 81

Macrotine................ 118
Magnesium, Mg.......... 57
  carbonate, MgCO$_3$ 54
  sulphate, MgSO$_4$....... 7
  mixture............... 8
Man — analysis of........ 165
Manganese, Mn.......... 51
Marsh's test............42-44
Mass — colored 26
Meconic acid............. 129
Meconidine 119
Mendelejeff's table...... 21
Mercuric salts............ 39
  chloride, HgCl$_2$....... 8
  iodide, HgI$_2$.......... 39
  nitrate, Hg(NO$_3$)$_2$..... 39
  oxide, HgO 39
  sulphide, HgS........ 39
Mercurous salts 35
  chloride, HgCl 8
  iodide, HgI 35
  nitrate, HgNO$_3$........9-35
Mercury, Hg...........35-39
Metals 18
Metathesis 19
Microcosmic salt......... 29
Milk —
  analysis of............ 111
  ash of................. 113
  butter in.............. 112
  caseine in............. 112
  condensed............. 111

PAGE.

Milk (continued) —
  fat in................. 112
  reaction to litmus....... 110
  specific gravity........ 110
  taste of................ 110
Millon's reagent.......... 90
Molecule 18
Molecular weight........ 21
Molybdenum, Mo......... 65
Morphene.............118-127
Monobasic acid.......... 20
Moore's test............. 92
Mucus (urine)........... 80
Mulder's test............ 94
Murexide 86

Narceine 119
Narcotine 118
Nessler's test.........55-103
Nickel, Ni............... 51
Nicotine................ 118
Nitrates (water) ....,.... 106
Nitric acid, HNO$_3$ 72
  oxide, NO............. 3
  peroxide, NO$_2$ 3
Nitrites (water).......106-108
Nitrogenous bodies....... 86
Nitrous acid, HNO$_2$ 71
Non metals............. 18
Nordhausen acid......... 108
Normal salt............. 20
Note book, How to keep a. 22

Organic acids...........73-77
  matter (water)......... 103
Oxalic acid.............47-73
Oxidation 19-51
Oxides..............18-24-26

Papaverine.............. 118
Perchloric acid, HClO$_4$ ... 73

PAGE.

Perissads.......... .. 21
Pewter................. 60
Phosphates (urine)....... 91
Phosphoric acid, H_3PO_4 . 47
Phosphorus, P... 41
Phosphuretted hydrogen.. 41
Picric acid.......... 66
Pilocarpine............. 118
Plant bases, separation of. 161
Platinum, Pt.......... 36
  chloride, PtCl4........ 13
Plu mber's solder....... 60
Podophylline............ 118
Poisons—
  aconitine, $C_{30}H_{47}NO_7$ ... 132
  analysis of........120-125
  antimony, Sb........ 142
  arsenic, As............ 146
  atropine, $C_{17}H_{23}NO_3$ ... 133
  brucine, $C_{23}H_{26}N_2O_4$ ... 127
  caffeine, $C_8H_{10}N_4O_2$ .. 137
  codeine, $C_{18}H_{21}NO_3$ 129
  conine, $C_8H_{15}N$........ 135
  copper, Cu............ 153
  how they produce death. 154
  lead, Pb............... 150
  mercury, Hg.......... 144
  morphine, $C_{17}H_{19}NO_3$ .. 127
  narcotine, $C_{22}H_{23}NO_7$ .. 129
  nicotine, $C_{10}H_{14}N_2$ 134
  phosphorus, P. 143
  prussic acid, HCN...... 139
  quinine, $C_{20}H_{24}N_2O_2$ ... 130
  strychnine, $C_{12}H_{22}N_2O_5$ 125
  veratrine, $C_{32}H_{52}N_2O_8$. 131
  zinc, Zn............... 152
Potassium, K............. 58
  bichromate, $K_2Cr_2O_7$. 9
  chromate, K_2CrO_4 9
  chlorate, KClO3........ 10
  cyanide, KCN or KCy.. 70
  ferricyanide, K_3FeCy_6. 10
  ferrocyanide, K_4FeCy_6. 10
  hydrate, KHO.... 45
  iodide, KI............. 10

PAGE.

Potassium (continued)—
  mercuric iodide (Ness-
    ler's test)..........55-103
  metantimoniate, KSbO3 58
  nitrate, KNO3......... 11
  nitrite, KNO2 12
  permanganate, K_2Mn_4-
    O_8 103
  sulphate, $KHSO_4$ 176
  sulphate, K_2SO_4 ...13-176
  sulphocyanide, KCNS... 12
Problems................. 168
Prussian (Fe4(FeCy6)3)
  blue.........'... . 3
Prussic acid, HCN...... 70
Ptomaines................ 116

Qualitative analysis in the
  wet and dry ways 23
  of calculi........... 98
  of urine............118-130
Quantitative volumetric and
  gravimetric.. 23

Rachitis................. 61
Radicals 18
Reaction 20
Reagents.......... 1-17
Reagents for Alkaloids—
  iodine in iodide of potas-
    sium................ 138
  Marme's......... 138
  Mayer's................ 138
  potassio-bismuth iodide. 138
  Scheibler's............. 138
Recapitulation Lab. work. 23
Reinsch's test..........42-44
Rhodium, Rh 36
Rhœagenine. 119
Rhœandine................ 119
Rubidium, Rb............ 59

PAGE.
Sabadilline 118
Salts —
  defined 20
  acid 20
  basic 20
  double 20
  neutral 20
Santonine 118
Sarcosine 119
Selenium, Se ....: 25
Semi-metals 18
Separation of acids 63–77
  bases 32–58
  Co and Ni 51
Signs of evolution and precipitation 1–19
Silicates 64
Silicic acid, H4SiO4 67
Silver, Ag 34
  chloride, AgCl 34
  chromate, Ag2CrO4 103
  nitrate, AgNO3 16
  standards 60
Sodium, Na 58
  acetate, NaC2H3O2 ... 13
  carbonate, NaCO3 13
  hydrate, NaHO ....: 14
  hypochloride, NaClO.... 15
  phosphate, Na2HPO4 .. 15
  phospho molybdate..... 15
  sulphide, Na2S 16
  sulphate, Na2SO4 176
  sulphite, Na2SO3 16
Solanine 118
Solution 32
  of indigo 16
Specific gravity — urine... 82
Standard solutions —
  ammonia 103
  Biette 147
  bromide of sada 84
  Donovan's 147
  Fehling's 93
  ferrous ammonium sul-
    phate 105

PAGE.
Standard Solutions (con.) —
  Fowler's 147
  Harle's 147
  lime 106
  Nessler's 103
  permanganate of potash. 103
  silver nitrate 103
  soap 106
Stannic chloride, SnCl4... 46
Stannous chloride, SnCl2. 61
  sulphide, SnS 47
Strychnine 118–125
Sublimation 24
Succinic acid 75
Sugar... .... 95, 113
Sulphates in urine........ 94
Sulphur, S 65
Sulphuric acid, H2SO4... 65
Sulphurous acid, H2SO3. 67
Synthesis 19
Symbol 20

Table, atomic weights.... 21
Tantalum 47
Tartaric acid............ 73
Tellurium, Te........... 25
Tests —
  alcohol............... 155
  alkalies 58
  aluminum 50
  ammonia............... 55
  antimony 43
  arsenic 42
  barium............... 56
  boric acid............. 66
  cadmium............. 44
  carbonic anhydride.... 67
  chloroform 155
  chromium............. 49
  cobalt............... 50
  copper 40
  gold 36
  iron................. 49

PAGE.

Tests (continued) —
lead.................. 33
magnesium............ 57
manganese............ 51
mercury............. 35
nickel................ 51
oxygen................ 24
ozone 160
phosphoric acid....... 41
platinum............13, 36
silicic anhydride 29
sulphur..............25, 64
tin..................... 46
water.................. 100
thermal units.......... 175
titanium 29
tungsten..............29, 32
Thebaine................ 118
Theine................. 118
Theobromine............. 118
Tin, Sn..... 46
Titanium, Ti............ 47
Tribasic acid............ 20
Tungsten, W... 64
Type metal............... 60

Uranium, U........... 46-92
Urine—
albumen 89
analysis........ 96
bile.................. 95
chlorides............ 90
composition 80
general reaction 81
phosphates...... 91
sugar 92
sulphates 94

PAGE.

Urine (continued) —
urea 83
uric acid............. 85

Venadium, V............. 65
Veratrine, $C_{32}H_{52}N_2O_8$ 118-131

Water............. 100
ammonia, free......... 104
"   albuminoid .. 104
chlorine............... 103
constitution 101
crystallization 101
hardness.............. 106
nitrates 106
nitrites.............106-108
organic matter......... 103
Witherite, $CaCO_3$....... 55

Xanthine, $C_5H_4N_4O_2$... 119

Yttrium, Yt............. 21

Zettnow's chart.......... 61
Zinc, Zn........45-48-52-152
alloys............... 60
carbonate 152
chloride.............. 152
hydrate 52
sulphate 152
sulphide 52
Zirconium, Zr.......... 21

BIBLIOGRAPHY.

The following books can be obtained of the publisher, **A. H. SMYTHE, Columbus, Ohio** :

QUALITATIVE ANALYSIS.
BEILSTEIN, DOUGLAS and PRESCOTT, ELIOT and STORER, FRESENIUS, GALLOWAY, O'BRINE, THORPE, VALENTINE, WILLL'S TABLES.

BLOWPIPE ANALYSIS.
BRUSH, ELDERHORST, PLATTNER, PLYMPTON.

VOLUMETRIC ANALYSIS.
FLEISCHER, HART, SUTTON.

WATER ANALYSIS.
FRANKLAND, WANKLYN.

URINE ANALYSIS.
HOFFMAN and ULTZMAN, NEUBAUER and VOGEL, THUDICHUM.

TOXICOLOGY.
BLYTH, CHRISTISON, OTTO, REESE, TAYLOR, WOODMAN and TIDY, WORMLEY.

QUANTITATIVE ANALYSIS.
CAIRNS, CLASSEN, CROOKE'S SELECT METHODS, FRESENIUS, THORPE, WOHLER.

MEDICAL AND PHYSIOLOGICAL.
ATFIELD, BOWMAN, GAMGEE, HOPPE-SEYLER, LEHMANN, O'BRINE, SAUNDERSON, VAUGHAN.

ORGANIC ANALYSIS.
ALLEN, BLYTH, PRESCOTT.

GENERAL CHEMISTRY.
BLOXAM, FOWNE, MILLER (3 vol.), NORTON, ROSCOE and SCHORLEMMER, TIDY, WURTZ.

CHEMICAL PROBLEMS.
COOKE, FOYE, JONES, THORPE.

CHEMICAL PHILOSOPHY.
COOKE, HOFMANN, REMSEN, TILDEN.

DICTIONARIES OF CHEMISTRY.
WATTS, STORER, COOLEY, GAMELIN, FRESENIUS' ZEITSCHRIFT FUR ANALYTISCHE CHEMIE.

www.ingramcontent.com/pod-product-compliance
Lightning Source LLC
Chambersburg PA
CBHW021711210326
41599CB00013B/1609